# Dialogue on Early Childhood Science, Mathematics, and Technology Education



Based on papers commissioned for the Forum on Early Childhood Science, Mathematics, and Technology Education, February 6–8, 1998 ▮ Washington, DC.

ISBN 0–87168–629–5



This publication and the February 1998 Forum on Early Childhood Science, Mathematics, and Technology Education were supported by a grant from the National Science Foundation (grant number ESI 96189093). Any interpretations and conclusions are those of the authors and do not necessarily represent the views of the American Association for the Advancement of Science or the National Science Foundation.

Cover photos: Young boy playing: Cummins, Jim/FPG International, LLC.
Boy planting tree: Taposchaner, Jacob/FPG International, LLC.

Book design by Jennifer Anne Kolansky.

Printed in the United States of America.

# Contents

# Preface

To thrive in a world increasingly shaped by science and technology, our children and grandchildren must own—in their own hands and minds—a basic understanding of how the world works, how we have come to know what we know, and the abilities to learn useful new knowledge and skills. Literacy in science, mathematics, and technology is not an option for the future. But where and when do we start?

Recent educational research suggests that even very young children have the ability to comprehend their world from a scientific perspective. Some studies indicate that children as young as three years old may be capable of concept-based theoretical learning. New research on how the brain develops during these early years promises to help us understand how young children learn mathematics and science. But what do we make of these findings? And how do we put them to good use in pursuing our goal of literacy?

In February 1998 a multidisciplinary group of more than 100 experts gathered in Washington, DC for the Forum on Early Childhood Science, Mathematics, and Technology Education. At the request of the National Science Foundation (NSF), the nation's most accomplished educators, scholars, and researchers convened for three days to discuss how, when, and even if we should teach science, mathematics, and technology to pre-kindergarten children. This book, *Dialogue on Early Childhood Science, Mathematics, and Technology Education* is a product of that meeting.

*Dialogue on Early Childhood Science, Mathematics, and Technology Education* represents some of the latest thinking about early childhood science, mathematics, and technology education. It brings together 11 papers on wide-ranging topics commissioned by the American Association for the Advancement of Science (AAAS) for the forum. Among the intriguing ideas to emerge from these papers are the following:

- Children are capable of learning more than we had previously thought, though we don't know enough about child development, yet, to say what experiences every child should have.
- Mathematics and science are usually absent in early childhood education.
- Early childhood teachers and caregivers are often ill prepared to incorporate appropriate science, mathematics, or technology experiences into children's lives.
- The range of early childhood experiences is vast, and the resources for early childhood education are few and inequitably distributed.

The book also contains an extensive bibliography and list of resources for educators, parents, and education groups to use as they seek the best science, mathematics, and technology experiences for young children. We hope that this book both sustains the conversation begun at the forum and encourages more of the community to participate.

Many people contributed to the success of the forum and to this book. From NSF, Dr. Margaret Cozzens, former director of the Elementary, Secondary, and Informal Education Division, initiated the project; Dr. Janice Earle, Project 2061's program officer, helped identify the funding; Dr. Alverna Champion, program officer, and Dr. Patricia Kenny, director of NSF's spectacular new Child Development Center, worked closely with us in every aspect of both the forum and book. From AAAS, Mary Koppal, communications director, organized both the conference and this book; Natalie Nielson and Terry Handy skillfully edited the manuscript; Barbara Walthal and Tracy Gath compiled the extensive bibliography and resource list; Lester Matlock, project administrator, handled all of the logistics for the forum. I am grateful to all for their dedication and hard work. Finally, I thank all of the authors for their thoughtful contributions to this volume.

*George D. Nelson*
*Director, Project 2061*

# Perspectives

# Early Childhood Education in Science, Mathematics, and Technology: An NSTA Perspective

*Fred Johnson*

The National Science Teachers Association (NSTA) believes two issues must be considered regarding early childhood education. First, we must understand how and why young children learn. Second, we must identify programs and learning experiences that apply this understanding of early childhood learning to effectively meet young children's needs.

Current research on brain development emphasizes the importance of early stimulation in developing brain connections from birth. The Carnegie Task Force on Meeting the Needs of Young Children in 1994 issued a call for help in preparing children for learning when they enter school. This report states that "brain development is much more vulnerable to environmental influence than previously suspected and early environmental influence on brain development is long lasting." Neurobiology research regarding normal brain function is revealing more about how children learn (Markezich 1996).

"Learning windows"—optimal times for learning at particular developmental stages—should be used to enhance understanding of science, mathematics, and technology in young children. Research findings have strong implications for developing effective early childhood education programs because "rich experiences produce rich brains" (Nash 1997).

Piaget's theory of cognitive development was created in the 1920s, long before access to medical imaging technology and current brain research was available. The current national redirection of science and math teaching is grounded in this theory, which stresses the use of a teaching/learning cycle and explorations through the manipulation of objects and materials. "Developmentally appropriate practice"—a curriculum based on

*Fred Johnson served as president of the National Science Teachers Association from June 1997 through June 1998.*

what is known about young children—should drive instruction (Clark 1996).

Research on children's motivation to learn and their underachievement reveals that young children are full of curiosity and a passion for learning (Raffini 1993). If this passion changes from delight to drudgery, one in four of those students will leave school before graduating. A greater understanding of student motivation is needed, particularly as it relates to intrinsic and extrinsic rewards for learning.

Documentation and evaluation data on Head Start, Title I, and the Military Child Care System may reveal models for effective preschool education. Closing the developmental gap between preschool children who are mentally stimulated by their family and surroundings and those who are not stimulated should be a priority in preparing children for school. The importance of brain development and the opportunities for early childhood stimulation calls for well-designed preschool education for three-, four-, and five-year-olds. These programs may compensate for a child's lack of stimulation in the previous years and months; they may also enhance less than stimulating home environments.

Cultural diversity and children with special needs are a particular challenge for early childhood education as developmental milestones are attained on a different schedule and in a different manner. If all children are to reach their potential, they must all be included in our concerns when we design and provide high-quality preschool educational opportunities.

Financing issues are always a concern. Local educational programming that is funded by grants and that receives special community support seems to be most effective: The stakeholders have more invested in the success of these programs. Educators' top priority should be financial assistance for preschool programs that are working effectively to prepare students for learning.

As NSTA considers neurological research and its implications for preschool education, we recognize the need for making the most of these early childhood years through well-designed preschool programs that provide science, mathematics, and technology education.

## References

Clark, J. V. (1996). *Redirecting science education*. Corwin Press, Inc.

Markezich, A. (1996). *Learning windows and the child's brain*. SuperKids Educational Software Review. Knowledge Share LLC.

Nash, M. J. (1997). Fertile minds. *Time*, 149:5.

Raffini, J. P. (1993). *Winners without losers: Structures and strategies for increasing student motivation to learn*. Upper Saddle River, NJ: Prentice Hall.

# Toward a Research Agenda in Early Childhood Science, Mathematics, and Technology Education

*Alverna M. Champion*

"It takes a village to raise a child." This well-worn African adage is no less true today than it was when the thought was first made word. The American Association for the Advancement of Science, with funding from the National Science Foundation (NSF) and input from the Office of Educational Research and Improvement (OERI) of the Department of Education, brought the global village together in February of 1998. Mathematicians and scientists, researchers and practitioners, teachers and administrators, and policy makers from near and abroad gathered in Washington, DC, for a dynamic Forum on Early Childhood Science, Mathematics, and Technology Education. We came together as a unique community: each with different views, beliefs, and ideals but all with similar commitments to improving the lives, opportunities, and fates of young children aged 3 to 5.

*Alverna M. Champion is professor of mathematics at Grand Valley State University in Grand Rapids, MI.*

We left the forum ready to inform the world about science, mathematics, and technology education for young children. We left revitalized to teach those who heretofore have been categorized as un-teachable. We left eager to encourage the alienated. We left amenable to welcome, with open arms, the disenfranchised. We left equipped to release the shackles from those children who are tethered to labels such as "at-risk." We left as we had come, reciting the mantra, "All children can learn," but now we actually believed it.

With so much synergy at work, how can the National Science Foundation afford not to support projects carefully crafted to improve the teaching and learning of mathematics, science, and technology for young children? With the needs of young children gaining the spotlight at all levels of government, it is important that we seize the day. Submit a proposal! Consider the following possibilities.

Margaret "Midge" Cozzens, the former director of the Elementary, Secondary, and Informal Science Education (ESIE) Division of the National Science Foundation, observed, "We don't have a good synthesis of the research on early childhood education as it relates to mathematics, science and technology." The Division of Research, Evaluation, and Communications and ESIE offer opportunities for funding in the areas of research, evaluation, and assessment.

Hyman Field, acting director of ESIE, adds to Midge's statement by encouraging the submission of proposals that would enhance young children's learning of mathematics, science, and technology in settings outside the formal classroom. Proposals for increasing parental involvement are a priority.

NSF has recently funded five projects on the development of early childhood instructional materials. NSF, in funding a variety of projects, seeks to learn more about what works, under what conditions, and for whom. These projects are not intended to be exemplars; they are intended to serve as examples. Funding innovative and creative instructional materials development projects that focus on substantive mathematics, science, and technology is particularly important to NSF.

# Making Sense of the World

*Shirley Malcom*

If people are shaped by their experiences—and I believe they are—then all of my experiences as a scientist and as a parent have convinced me that science, mathematics, and technology matter *for* young children because they matter *to* young children. As children set the table, match their socks, or reach for their jackets, boots, and mittens on a snowy day; as they learn to cook or play with bath toys, they amass experiences that set them up for the punch line. And the punch line is this: We live in a world that is governed by rules, where some outcomes are predictable, where knowledge can be uncovered, where questions can be asked and answered.

My own children taught me early on that they sought—and would create if necessary—explanations of their world and its workings. They took no part of the material world for granted. What makes it snow? What are clouds made of? My daughter Kelly's fear of thunder led her to a moment of scientific theorizing when she was four years old. As we were racing home during a storm, she announced from the back seat that she knew why the "clouds bumped into each other and make that thunder." "Why is that?" I asked. "Because they don't have any eyes," Kelly replied.

Child development research tells us that children do, in fact, attribute the characteristics of animate objects to natural phenomena. But what was most fascinating to me was that she found it necessary to articulate a hypothesis. Later experiences would help her refine and develop her theories about thunder, storms, and clouds, but for now she had asked a question that

---

*Shirley Malcom is director of Education and Human Resources at the American Association for the Advancement of Science.*

was important to her and had formulated an explanation that would help her make sense of the world.

Around the time my children began to confront me with their questions and hypotheses, I was asked to make a presentation on science, mathematics, and technology at a meeting of the National Association for the Education of Young Children. To prepare, I began to explore the formal knowledge about child development and the early childhood years in particular. At the same time, I reflected on what I had learned from watching my own and other young children frame their questions about the world and develop their hypotheses. It became clear to me that by their very nature, science and mathematics could offer young children powerful ways of knowing about the world. And that indeed, children's experiences in their very early years could prepare them for the formal study of science, mathematics, and technology later on. These conclusions led me inevitably to my present role as advocate. My message is this: We need to provide *all* children with much greater access to the richest variety of experiences that will help them make sense of their world. We must have unbounded expectations for every child.

## The Great Equalizers

Some may fear that I am suggesting we impose a rigid, formal curriculum on young children. That is not at all what I have in mind. Instead, we must take advantage of children's everyday experiences, supplementing them with other experiences that are consistent with what we know about how children develop and what we know about each individual child. The value of such an approach was best described by David Hawkins in an article in the Spring 1983 issue of *Daedalus*:

> "The kind of experiential background in children's lives before schooling begins or along the way is more uniformly adequate to math and science than to most other school subjects. The poverty or richness of social background matter less here in the early years than in other school subjects. Math and science should therefore be the great equalizers, whether they are now seen to be or not."

Unfortunately, mathematics and science are often the great separators rather than equalizers. They are the gatekeeper subjects. Without advanced courses in science and mathematics, students are excluded from educational opportunities and experiences that can affect their career aspirations, their role in society, and even their sense of personal fulfillment. With this much at stake, how do we make sure that all young children have the important experiences that will provide them with a strong foundation for future learning? At the very least, it will require much more cooperation among three distinct communities that are just beginning to take account of each other.

## Building the Bridges

For those in the K–16 education community, thinking about science, mathematics, and technology in a preschool context is relatively new. Similarly, the early childhood education community is just beginning to consider what content-rich programs would be like at the preschool level. And for those who work in the area of educational equity, there is now a growing awareness that access to thoughtful, engaging experiences in science, mathematics, and technology during the early childhood years can provide both short- and long-term benefits to all children. Connecting these three communities and finding out where their interests intersect are essential first steps.

In the area of informal education—television, museums, science centers, and the like—some programs and initiatives have already begun to make these connections. Museum programs aimed at young children have dealt with a variety of science, mathematics, and technology concepts: hot and cold, big and small, sink and float, machines, numbers, senses, shapes, and so on. Unfortunately, many children have not had an opportunity to take part in these kinds of programs. Much more needs to be done to make such programs accessible to all families.

Television—with its almost universal accessibility—along with toys, games, computer software, films, and books have also had some success in exposing young children to science and mathematics. These few examples serve only to suggest the range of activities, materials, and media that can contribute to a child's exploration and understanding of the world.

In 1985, the *Science Teacher* published an article about holistic learning in science by Bob Samples and Bill Hammond. Whether the authors knew it or not, they were actually asking for an inclusion model for school science. That model drew heavily on models that are the norm in early childhood education, where it is widely accepted that each child develops at an individual pace and gains skills and knowledge in a distinct and personal style. The authors were negative about book-bound, lecture-based instruction, and they made this statement: "Learning style researchers are convinced that at any given time half to three-quarters of the students in most classrooms are not learning at near optimum."

I think it is probably higher than that. Samples and Hammond go on to say, "Only a quarter of our students gain the most from a single approach to instruction. We must try more effective ways to reach all students. It is a kind of myopia to cling to instructional practices that systematically exclude students. It is a kind of immorality to support education practices that systematically exclude most of a learner's mind." There is much we can learn from our colleagues in early childhood education.

## What Works?

In the mid-1980s the American Association for the Advancement of Science (AAAS) collaborated with the National Urban League on a project funded by the National Science Foundation. The goal of the project was to develop model programs for early childhood education in science that could be customized for preschool settings. The models had these constraints: few of the preschool staff had four-year degrees in early childhood education; many had associate degrees; the staff turnover rate was high, as it is in most programs for young children. In other words, we had to design a model that would respond to the realities of the world as it was, not as we wanted it to be. Our challenge was to help the people who were working with very young children to become empowered themselves by their own knowledge of science and the fact that it had meaning in their lives. Until that happened, they were unlikely to engage the children in science, mathematics, and technology.

These models have also been an influence on AAAS's Black Churches project. We have found that early childhood education and childcare programs are among those programs most frequently offered by the churches. How do we help churches improve the quality of what they are doing? How do we help their staff develop themselves professionally? We have found once again that it is a matter of enabling individuals to take hold of the science themselves. We engage them with the kinds of strategies that actually work with young children and then show them the science, mathematics, and technology that is present in the early childhood environments they have created. Chances are, there is already a corner for blocks or perhaps a water table in the space. By using these to construct towers, build a big block from several smaller blocks, pour water into containers of different sizes and shapes, or predict how far a ball or toy car will roll on different surfaces, children are able to interact with and learn about the natural world.

Several years ago, I served as co-chair of the Carnegie Task Force on Learning in the Primary Grades. We examined many issues related to children and their well-being, but a pivotal finding of our work was this: For three-year-old children—whatever their situations in life—everything was still possible. But in just a few years time, many would begin to lose ground. How do we support that amazing potential of early childhood? What expectations and outcomes should we have in mind when we talk about preschool education? How specific need we be in setting goals for content and programs? How do we maintain a focus on the development of each individual child while ensuring that all children have access to the kinds of experiences that will prepare them for high achievement in science, mathematics, and technology? Unless we name those experiences, some children will simply not get them.

One goal of the Forum on Early Childhood Science, Mathematics, and Technology Education was to name those experiences. What is it that students need to know and be able to do by the time they get to kindergarten? What experiences can we assume they have had? How can we help provide an opportunity for every child to have those experiences?

## Lifting Refrigerators and Learning about Levers

Finally, a story of my other daughter illustrates where such thinking might lead. My husband needed to move the rollers under the refrigerator. He asked our five-year-old for help. "Lindsey," he said, "I want you to lift the refrigerator." Lindsey turned to her father, laughed, and said, "Daddy, you're so silly. I'm too little. I can't lift a refrigerator. It is too heavy for me."

Her father went out to the garage and came back with a long-handled shovel. He placed the blade securely under the front of the refrigerator and pushed down on the handle. He called Lindsey over and as they gradually exchanged weight and she put her 40 pounds on the shovel handle, she lifted the refrigerator. Her eyes lit up, her mouth dropped open, and this silly grin came across her face. She had discovered that technology was empowering. We never said the word "lever" to her. We never tried to tell her about fulcrums or anything else. She had found out herself.

When the nation's governors and President Bush met in Charlottesville for the 1989 Education Summit, the first national education goal they articulated was this: *All* children will come to school ready to learn. It has taken us many years to learn what "ready" means. It means that children will come to school with loving and caring adults in their lives, with adequate health care and nutrition, and with an entire community of support. But it also means that they must come to school with the experiences that will allow them to reach their full potential and with limitless expectations that they will succeed at the highest levels.

## References

Hammond, B. and Samples, B. (November 1985). *The Science Teacher*, 52(8):40–43.

Hawkins, D. (Spring 1983). Nature Closely Observed. *Daedalus*, 112(2).

# The Forum on Early Childhood Science, Mathematics, and Technology Education

*Jacqueline R. Johnson*

The forum on Early Childhood Science, Mathematics, and Technology Education, convened by Project 2061 of the American Association for the Advancement of Science, created a learning community of early childhood practitioners and researchers, scholars, and technological experts from the sciences and mathematics. These individuals explored the status of mathematics, science, and technological education in the early childhood years. The forum was convened in February 1998 in an effort to:

- merge three constituencies—representatives from the mathematics, science, and technology communities; early childhood educational practitioners; and the educational equity community—to enhance mathematics, science, and technology education in early childhood.
- consider what the goals of such education should be and to begin to articulate strategies for achieving these goals.
- review what is known about this area of education.
- identify promising subject areas or programs worthy of outside funding.

*Jacqueline R. Johnson is professor of sociology and chair of the department of anthropology and sociology at Grand Valley State University in Allendale, MI.*

This paper synthesizes the issues and findings from the forum. It provides an overview of what we know about mathematics, science, and technology education; identifies exemplars of good practice; and identifies obstacles to goal achievement. Specifically, this paper considers the issues and agenda that prompted the meeting. It also presents a preliminary agenda for future work and possible funding initiatives in this arena.

In "Science Assessment in Early Childhood Programs" (page 106), Edward Chittenden and Jacqueline Jones recount a kindergarten class's observation of a dead fish. The narrative provides an example of the promise and potential of early childhood science education. In "Young Children and Technology," Douglas Clements likens the power of effective scientific and mathematical thinking (which he describes as "integrated-concrete thinking,") to the strength of sidewalk concrete (page 100). In each case, strength is provided by "the combination of many separate ideas (particles) in an interconnected structure of knowledge." These two examples, which I have labeled "dead fish" and "sidewalk glue," respectively, aptly characterize the ongoing work of practitioners and researchers in early childhood education.

## Why Now?

Three main factors provided the impetus for the Forum on Early Childhood Education in Mathematics, Science, and Technology: (1) increasing numbers of children are enrolled in some type of preschool program; (2) widespread agreement exists on the need for students to be science literate if they are to succeed in today's rapidly changing world, yet few preschool programs address science, mathematics, and technology; and (3) the growing number of new technologies opens up myriad possibilities for preschool learning. These factors are discussed in the following sections.

### *Children Are Not at Home Anymore*

As one presenter stated in her introductory remarks, "This forum is in large part a response to reality—children ages three to six aren't at home anymore." Indeed, the percentage of children three to five years of age enrolled in nursery school or preschool programs in each state ranges from 50 percent to 69 percent. Although the United States lags far behind other industrialized democratic countries in its provision of public pre-primary education for three- to five-year-olds, the actual number of children in preschool daycare or educational settings has grown dramatically in the past 10 years. Increased numbers of single parents and the workforce participation of both parents in two-parent households are among the factors contributing to this tremendous growth.

State spending for these programs varies from $0.24 per child in Idaho to more than $70 per child in Alaska, Connecticut, Massachusetts, New York, and Vermont. Predictably, the difference in the types of care experienced by preschool children is considerable. For some families, preschool care takes place in an educational setting with highly qualified professionals in charge. For many others, however, the pre-primary experience takes place in daycare centers or private homes—licensed or unlicensed—

where workers are poorly paid and largely untrained and where worker turnover is high.

It is clear that the education of preschool-age children is no longer solely a family, private endeavor. With increasing numbers of preschool children in "public" settings for extended periods of time, there is a need to create some coherence in what children learn. Yet high-quality preschool education is far from commonplace—even less common are preschool programs that include science, mathematics, and technology. The welfare of our children is of concern to us all as they are the future of a democratic society. This concern was one of the principle reasons this forum was held.

### *The Literacy People Got There First*

Conference participants carefully documented the long-term benefits—educational and social—of high-quality preschool education. In part, shifts in this country's social, political, and economic systems have resulted in more children being "available" as potential beneficiaries of high-quality preschool programs. For a variety of reasons, however, even high-quality programs do not emphasize mathematical, scientific, and technological learning.

Education in mathematics and science has long been given minimal attention in early childhood education programs. There is a sense that the literacy people got there first. In other words, there is a common perception that the language arts play a predominant role in early childhood education, to the exclusion of mathematics and science. The forum was convened in part to address this issue. It focused on the need to place science and mathematics education in the preschool curriculum and on the need to consider the role that each subject plays in the cognitive and intellectual development of the child.

Recent findings regarding the relatively poor international performance of American students in mathematics and science have heightened concerns and raised anew questions about the "preparedness" and "readiness" of our young people as they enter school. Have educators, communities, and parents done all they can do to set the stage for success in mathematics, science, and technology at the earliest levels of formal education, that is, at preschool?

These concerns allowed participants in the forum to articulate a number of questions related to the topic of science, mathematics, and technology education in early childhood. What do we know about the predisposition of young people to learn science and mathematics concepts and theories between the ages of three and eight? What might we need to "unlearn"? How do we measure the "success" of our efforts? How do we import and export models of success to arenas other than those in which they are piloted? These questions, raised against the backdrop of our social and economic concerns with intellectual competitiveness in an international arena, were another motivating force behind the forum.

### *The Seductive Computer*

A third factor that prompted this conference relates to the transformation of intellectual inquiry brought about by the ready availability of computer and other information technologies. In documenting the availability of computers in preschool settings, Douglas Clements states that the computer to student ratio changed from 1:125 in 1984 to 1:10 in 1997 (page 92). The question is no longer whether computers can assist in learning in general, and in mathematical and scientific learning in particular, but rather how computers can and should be used and to what end. The presence of new technologies—including child-friendly computer software in mathematics—and the ease with which young children use these technologies, prompted us all to consider the various ways that computers can foster the skills, the conceptual frameworks, and the social aspects of mathematics and science learning.

## What Do We Know?

One important outcome of this forum is that it helped to identify and illuminate what the varied constituencies already know about several seemingly disparate areas, each of which significantly affects early childhood education.

### *Rethinking Children's Conceptual and Theoretical Abilities*

David Elkind points out in "Educating Young Children in Math, Science, and Technology" (page 62) that mathematics, science, and technology are adult abstractions. Nevertheless, the world of the child is, in fact, replete with opportunities to directly experience these abstractions. Moreover, children are always engaged in efforts to make sense of the world. Science is a way of thinking, a way of being in the world. The challenge is to find the means to encourage mathematical and scientific thinking in such a way that enhances children's readiness to learn and expands children's concepts and knowledge. As Susan Sperry Smith suggests in "Early Childhood Mathematics," (page 87) the teacher's task is to help children "re-invent" their knowledge of science and mathematics.

Although in the past educational research has cast doubt on the abilities of very young children to understand science and mathematics, more recent work grounded in developmental and cognitive psychology suggests that children are indeed capable of concept-based theoretical learning. In "Concept Development in Preschool Children," (page 50) Susan Gelman carefully documents this new approach. She identifies four key themes from recent research:

- Concepts are tools and as such have powerful implications—both positive and negative—for children's reasoning.
- Children's early concepts are not necessarily concrete or perceptually based. Even preschool children are capable of reasoning about non-obvious, subtle, and abstract concepts.
- Children's concepts are not uniform across content areas, across individuals, or across tasks.
- Children's concepts reflect their emerging "theories" about the world. To the extent that children's theories are inaccurate, their conceptions are also biased.

In addition, Gelman calls our attention to the difference between what children actually do and what children can do, i.e., their capabilities. Similar to other contributors to this forum (especially New and Clements), Gelman cautions us to avoid underestimating the cognitive, conceptual, and theoretical abilities of the child. Indeed, she claims that the purpose of her paper is to shatter "several myths about children's early concepts" (page 57). She provides evidence that shows "that even preschool children make use of concepts to expand knowledge via inductive inferences, that children's concepts are heterogeneous and do not undergo qualitative shifts during development, and that children's concepts incorporate non-perceptual elements from a young age.... Children's concepts are in fact far more sophisticated than has been traditionally assumed..." (page 57).

Douglas Clements raises similar issues in relation to the potential of the computer—a tool that facilitates abstract and higher order reasoning—to enhance the cognitive and intellectual development of very young children. (See page 92.) This is particularly effective in relation to mathematics reasoning. Clements illuminates the manner in which children use the computer to engage in complex computational activities previously thought to be beyond their reach. Depending on the spatial arrangements in which computers are housed and the manner in which they are used in the classroom, computers can also facilitate the collaborative and cooperative process that is the foundation of successful scientific reasoning and inquiry. Clements, however, in spite of—or perhaps because of—his belief in the educational effectiveness of computer technology, cautions us that technology serves the science and art of teaching and learning, not the other way around.

Thus, the forum revealed in a public way the findings of contemporary researchers: We know that children are capable of more than we had previously thought. We know that, given the resources, we can provide very young children with inviting technological products that will further develop their abilities. Yet we also know that mathematics and science are too often absent in early childhood settings.

### *Rethinking Early Childhood Curriculum*

Mathematics and science education are given short shrift in the preschool context. In addition, education in mathematics and science at this age level seems counterintuitive to many. Science and mathematics have been perceived and presented as too formal, too abstract, and too theoretical—in short, too hard for very young children and their teachers. Moreover, the "constructivist" approach in educational philosophy, which places the child at the center of the educational process and the teacher in the non-authoritarian role of observer, facilitator, even "outsider," may fuel the misconception that science and mathematics education should not occur at the early childhood levels. Science instruction has long been teacher-centered, with the teacher as the authority figure. However, early childhood educators know that authority-based, teacher-centered instruction is inappropriate for preschool children. While research on effective teaching and learning refutes the notion of teacher-as-authority and new approaches to science education eschew it, the image persists. This perception has led many educators to abandon the concept of teaching science to young children. One of our challenges, then, is to rethink both science education and preschool curriculum.

### *Teaching the Teachers*

Those least likely to be educated in mathematics and science are ironically the ones most likely to be in the classroom or in child-care settings. Not only is there an extraordinary range in educational background of early childcare workers, certified teachers often express an aversion to mathematics and science and are concerned about their abilities to instruct in these areas. The paper by Juanita Copley and Yolanda Padrón addresses this issue specifically, noting that "few professional development programs focus specifically on mathematics and science concepts in early childhood" (page 120).

Forum participants discussed this issue at length and concluded that—at least for certified teachers—general science courses in colleges and universities need to be redesigned with mathematics and science literacy in mind. At my own institution, a National Science Foundation project funded earlier in this decade led to the creation of general education classes with special hands-on lab components for prospective elementary teachers. These courses in biology, geology, physics, and chemistry epitomize the ideas expressed in the forum—that science is a way of being, thinking, and doing. Interestingly, these classes have become almost as attractive to non-education majors as they are to those seeking certification, and not because they are easy. Rather, they enliven science in a way that more traditionally designed science classes have been unable to do.

### *Good Practice*

Several presenters from the forum remind us that there are many long-standing models of good practice in early childhood education, including Montessori, Highscope, Creative Curriculum, and Reggio Emilia. For the most part, these approaches emphasize the role of the teacher as facilitator, provider of appropriate context, provocateur. As one of the presenters said, these models encourage children to get "stuck," then help them to discover ways to get "unstuck." The models are grounded in the belief that children learn best by doing. Within this framework, "children are encouraged to handle objects, observe and predict results, hear and use language, and collaborate with adults and older children to develop ideas." This approach seems especially conducive to learning scientific "habits of mind," even as it calls into question more traditional conceptions about science education, which place the teacher in the role of authority.

Implicitly or explicitly, all the models emphasize the importance of observing and documenting children's work—including documenting and recording children's "talk"—as a way of discovering how children learn best. "Science Assessment in Early Childhood Programs" by Chittenden and Jones identifies a variety of measures that can assess the effectiveness of programs for mathematics and science education and advocates, as most assessment experts do, the use of multiple measures.

In "Early Childhood Mathematics," Susan Sperry Smith provides further models of good practice in early childhood mathematics and science education. Smith takes us into the world of the classroom and provides us with tangible examples of how early mathematics education can "work" appropriately and effectively in the hands of knowledgeable and accomplished teachers.

## Barriers and Obstacles

We have seen that there are examples of good practice. There are also emerging paradigms that call into question previously held beliefs about the limited abilities of very young children in terms of their conceptual and theoretical sophistication with respect to science and mathematics. New computer software further encourages scientific thinking and doing—reflection and synthesis, collaboration and cooperation. Nevertheless, the excitement generated by the promise of this research and these models is tempered somewhat by the reality of making them available to all of our nation's children. What are the other obstacles to our success?

### *Teacher Preparation*

First we need to teach the teachers. Mathematics, science, and technological education of our preschoolers will not take place unless teachers are appropriately disengaged of their fears and anxieties in these areas.

### *Reproducing Social Inequalities in Educational Programs*

Effective education cannot occur as long as access to high-quality preschool educational programs remains inequitable. Even though research indicates the need to invest in early childhood education and illustrates the benefits accrued when we do, "...Only 45 percent of three- to five-year-olds from low-income families [are] enrolled in early childhood programs, compared with 71 percent from their high-income counterparts" (Day and Yarbrough, page 32–33). Statistics from the Annie E. Casey Foundation cited on page 33 "reveal that a large portion of children do not participate in early childhood programs." Furthermore, "...the range of financial investment in early childhood education varies greatly from state to state" (Day and Yarbrough, page 33). Keep in mind that these data only pertain to those children who are actually in relatively formalized early childhood educational programs. They do not account for those children enrolled in daycare programs—licensed or unlicensed—that lack any structured educational component.

New reminds us on pages 141–142 that education, albeit inadvertently, reproduces social and cultural systems, including the existing system of economic inequity. On pages 118–119, Copley and Padrón argue that these inequities are further exacerbated by the increasingly diverse cultural and language backgrounds of children in the American educational system, the lack of teacher preparedness to teach in the face of this diversity, and teachers' lack of facility with mathematics and science. One has to wonder where to begin.

Our sensitivity to diversity may actually perpetuate educational inequities. In our efforts to be fair, to acknowledge multiple cultural backgrounds, and to respect difference, we may abdicate our responsibilities to create common bodies of knowledge and abilities, thus closing doors well before high school to those who most need them opened (see New, page 143). Science and mathematics, because of their ubiquitous presence in the daily experiences of all groups, ought to be the great equalizers. Instead, they have long been the great dividers. The need to reverse this trend is apparent, and more pressing than ever.

Providing access to high-quality early childhood education programs, finding ways to successfully replicate pilot programs that work to reach children on a larger scale, and overcoming the anxieties that teachers themselves feel toward the subject present the greatest obstacles to success in this domain. These areas should receive considerable attention and resources in the near future.

## Support for New Initiatives

What needs to be done? What kinds of initiatives deserve support? Several possibilities emerged from the work of this forum.

### *Materials Development and Dissemination*

Although many of the resources needed for science and mathematics education are available in the natural environment of the child, scores of existing curriculum materials, mathematics manipulatives, and computer software enhance this learning. All children would benefit from efforts to disseminate these materials more widely. The U.S. Department of Education's Educational Resources Information Center (ERIC) already provides a resource bank for teachers and parents. Linking early childhood personnel to existing resources, increasing their familiarity with the materials available, and increasing the technological literacy of early childhood practitioners are all worthy endeavors. Encouraging practitioners to use online data sources and to communicate with their peers via online study groups could significantly enhance preschool efforts in mathematics and science literacy,

as well as increase the technological literacy of those same practitioners.

### *Collaborations Between Researchers and Practitioners*

Practitioners attending the forum warned against the dangers of putting the research scientists in charge. Instead, they suggested that providing opportunities for academics and early childcare employees to work side-by-side could produce the most fruitful results. Beginning on page 123, Copley and Padrón recount the successes of several existing programs that engage early elementary school educators, prospective teachers, and university faculty in what appear to be highly productive exchanges. We should attend to the practitioners' warnings and further explore and foster such "side-by-side" initiatives.

### *Science and Mathematics General Education*

General education initiatives that can help prospective early childhood educators to overcome their anxieties about science and mathematics are worthy of support. Nothing is more apparent from this forum than the need to enhance the mathematics and science education of preschool teachers. Their preparation in these subjects is the cornerstone of children's success. If the teachers see themselves as unable, if they believe that mathematics and science are too hard, then we will never reach the first national Educational Goal as defined by Congress and the nation's governors: "All children come to school ready to learn."

### *Science and Assessment*

We would do well to support research efforts that 1) examine the implications of K-12 standards in science and mathematics for early childhood education and 2) determine appropriate ways to assess our efforts to impart this content to very young children.

## Final Words

Given the gross inequities in the provision of high-quality early childhood education, initiatives that benefit the most needy families must receive top funding priority. Even with the best assessment and evaluation measures, it is difficult to sort out the reasons for the success of initiatives and programs that occur in privileged settings. Numerous advantages exist in these settings—in the physical environment, the superior education of the personnel, the teacher/student ratio, and even in the prior experiences of the children themselves. While it is encouraging that some programs work, we will only enjoy true success when these programs reach across economic and social lines to serve teachers and children in the most disadvantaged circumstances.

## References

Bowman, B. (1998). Policy Implications for Math, science, and technology in early childhood education. *Dialogue on early childhood science, mathematics, and technology education.* Washington, DC: Project 2061, American Association for the Advancement of Science.

Chittenden, E., and Jones, J. (1998). Science assessment in early childhood programs. *Dialogue on early childhood science, mathematics, and technology education.* Washington, DC: Project 2061, American Association for the Advancement of Science.

Clements, D. (1998). Young children and technology. *Dialogue on early childhood science, mathematics, and technology education.* Washington, DC: Project 2061, American Association for the Advancement of Science.

Copley, J.V., and Padrón, Y. (1998). Preparing teachers of young learners: Professional development of early childhood teachers in mathematics and science. *Dialogue on early childhood science, mathematics, and technology education.* Washington, DC: Project 2061, American Association for the Advancement of Science.

Day, B., and Yarbrough, T. (1998). The state of early childhood programs in America: Challenges for the new millenium. *Dialogue on early childhood education: Expert research and views on when and how children should learn science, mathematics and technology.* Washington, DC: Project 2061, American Association for the Advancement of Science.

Elkind, D. (1998). Educating young children in math, science, and technology. *Dialogue on early childhood science, mathematics, and technology education.* Washington, DC: Project 2061, American Association for the Advancement of Science.

Gelman, S.A. (1998). Concept development in preschool children. *Dialogue on early childhood science, mathematics, and technology education.* Washington, DC: Project 2061, American Association for the Advancement of Science.

New, R.S. (1998). Playing fair and square: Issues of equity in preschool mathematics, science, and technology. *Dialogue on early childhood science, mathematics, and technology education.*

Washington, DC: Project 2061, American Association for the Advancement of Science.

Smith, S. S. (1998). Early childhood mathematics. *Dialogue on early childhood science, mathematics, and technology education.* Washington, DC: Project 2061, American Association for the Advancement of Science.

# A Context for Learning

# The State of Early Childhood Programs in America: Challenges for the New Millenium

*Barbara Day and Tracie Yarbrough*

## Where Are the Children And Who's Minding Them?

When ranked among other industrialized countries, America comes in first in many areas. Among these are military technology and exports, Gross Domestic Product, the number of millionaires and billionaires, health technology, and defense expenditures. When ranking the factors related to the welfare of our children, however, America does not rate so high. In the income gap between rich and poor children and in infant mortality, America ranks 18th; 17th in efforts to lift children out of poverty and in its rate of children born with low-birth weights; and 16th in living standards among the poorest one-fifth of children. In addition, America ranks below average in math scores among 41 countries and last in protecting our children against gun violence (Children's Defense Fund [CDF] 1997). According to the Center for Disease Control and Prevention, the death rates of American children under 15 as a result of firearms are far higher than the combined rates of 25 other industrialized countries. Three out of every four children murdered in the 26 countries combined were American children (CDF 1997).

There are a variety of circumstances that not only place our youngsters in physical danger, but contribute to the fact that many begin school ill-prepared to learn. According to the National Education Goals Panel (1993), several factors have placed about one-half of our youth at risk of school failure. Among these are insufficient opportunities to develop mentally and physically, as almost 25 percent of all mothers in this country receive little or no prenatal care. In addition, larger numbers

*Barbara Day is professor and chair, curriculum and instruction, at the University of North Carolina–Chapel Hill. Tracie Yarbrough is a doctoral assistant in curriculum and instruction at the University of North Carolina–Chapel Hill.*

of single-parent families have placed strain on the family unit, which translates into some children having problems at school.

In March 1997, the Children's Defense Fund released 20 key facts about American children (CDF 1997). These statistics are illustrative of the severe problems facing the youth of America, including:

- 1 in 2 will live with a single parent at some point in childhood,
- 1 in 4 is born poor,
- 1 in 4 is born to a mother who did not graduate from high school,
- 1 in 5 lives in a family who receives food stamps,
- 1 in 7 has no health insurance,
- 1 in 8 is born to a teen mother,
- 1 in 12 has a disability,
- 1 in 14 was born at low birth weight,
- 1 in 21 is born to a mother who received late or no prenatal care, and
- 1 in 25 lives with neither parent.

The Children's Defense Fund urges citizens to take action in light of these staggering facts about America's youth. Other statistics reported in the *CDF Yearbook* include the inexcusable fact that an American child drops out of school every eight seconds, is reported neglected or abused every 10 seconds, and is arrested every 15 seconds (CDF 1997).

## The Need for Early Childhood Education

These statistics should create concern for all citizens, as the welfare of our children affects the future of us all. Something must be done to protect our youth from such misfortune. In order to combat such problems, large investments in education must be made, particularly at the early childhood level. A strong education in the early years can have a dramatic effect on a child's life and well-being. Experts in early childhood education tell us that "in the first three years of life, children learn, or fail to learn, how to get along with others, how to resolve disputes peaceably, how to use language as a tool of learning and persuasion, and how to explore the world without fear" (CDF 1997). In addition, brain research reveals that most of the connections that will be maintained throughout life are formed during childhood (Bredekamp and Copple 1997). The Packard Foundation reported that children who participate in high-quality programs in the early years are less likely to need remedial education later on and are less likely to participate in acts of juvenile delinquency (CDF 1997).

Probably the most notable study of the benefits of early childhood education is the High/Scope Perry Preschool Project. This long-term study of 123 disadvantaged black youths began in 1962 and is continuing today. At ages three and four, these children were divided into two groups, one group receiving high-quality preschool education and the other receiving no preschool education at all. A number of variables were studied, including the children's abilities, attitudes, academic achievement, involvement in criminal behavior, participation in welfare programs, and patterns of employability. The results through age 19 showed that those who received a high-quality preschool education completed high school at a higher rate, attended college or job-training programs more frequently, held down jobs at an increased rate, were arrested fewer times for criminal acts, and needed public assistance less frequently than those who did not receive such education (Weikart 1989). A recent assessment of the students at age 27 concluded that the children who participated in the high-quality preschool program had fewer criminal arrests and earned more money than their disadvantaged counterparts. It is estimated that over these students' lives, $7.16 is saved for each dollar invested in preschool education (Smith et al. 1995).

The Cost, Quality, and Child Outcomes in Childcare Centers study (1995) found that "children in higher quality preschool classrooms display greater receptive language ability and pre-mathematics skills, and have more advanced social skills than those in lower quality classrooms." The study also revealed that children in high-quality preschool programs "have more positive self-perceptions and attitudes toward their childcare, and their teachers are more likely to have warm, open relationships with them." These factors all contribute to increased readiness for school (Cost, Quality, and Child Outcomes in Childcare Centers 1995).

Aside from preparing children for future school years, well-documented research indicates that a number of other benefits can be reaped from high-quality early educational experiences (Bridgman 1989; Smith et al. 1995):

- A safe and caring environment is provided for children.
- Social problems such as violence and delinquency among juveniles, teenage pregnancy, welfare dependence, and school failure are prevented.
- Good health and nutritional practices are promoted.
- Children's social, physical, emotional, cognitive, and psychological development is fostered.
- Families are strengthened.
- Welfare recipients are given an opportunity to work, therefore becoming more self-sufficient.

A strong educational beginning for young children is not only of concern to educators and parents, but to those in business and politics as well. Research indicates that "early childhood education is critical to the nation's future economic position because it provides members of the next generation of workers with a solid foundation of skills, competencies, attitudes and behaviors that will ensure their success in a more technology-based and competitive future economic environment" (Smith et. al 1995). In his 1992 address to the Nation's Governors, President Bush stated that "by the year 2000, all children in America will start to school ready to learn." In fact, one initiative that was central to the Goals for America's Children issued by the National Education Goals Panel was identifying the need to increase this country's investment in high-quality early childhood education (Seefeldt and Galper 1998).

## Barriers to High-Quality Experiences

### *Poverty*

Many of the problems facing today's children are a result of poverty. In 1994 it was reported that approximately 15 million U.S. children lived in poverty, at that time the largest number then in almost three decades. It is estimated that each year of poverty for these children costs society between $36 million and $177 billion dollars in lower productivity and employment (Sherman 1994).

Children who are born poor are at a greater risk of educational failure. According to Bowman (1994), most poor and minority children are at risk for developmental failure. This problem is exacerbated by the conflict that exists between the behaviors valued at home and those valued by the school, she contends.

Children born to poor families are burdened with inadequate resources as well as the many other problems associated with poverty. It is estimated that a year of childcare for just one young child can cost a family $4,000 (CDF 1997). The childcare that poor parents can obtain is inadequate at best. Many of the childcare centers serving the poor function as babysitting services rather than instructional institutions.

### *Participation Rates*

As if low quality among educational programs during these very impressionable years is not troublesome enough, some youth never have the opportunity to participate in early childhood programs at all. If the lack of money does not prohibit them from attending, long waiting lists do. In general, students who need the help the most are the ones who cannot afford it. Consequently, studies showed that only 45 percent of three- to five-year-olds from low-income families were enrolled in early childhood programs, compared with 71 percent from their high-income counterparts (CDF 1997).

Table 1 shows the percentage of three- to five-year-olds enrolled in nursery school or kindergarten in 1993, as determined by the Annie E. Casey Foundation (1997). These statistics reveal that a large portion of children do not participate in early childhood programs.

All 50 states seem to be fairly comparable in the number of children receiving childcare services. The range from the smallest percentage to the largest is 22 points, and all but four states have between 50 percent and 69 percent of these children enrolled in such programs. This phenomenon may be due in part to the passage of the Family Support Act and At-Risk Childcare Legislation, which requires that states match funds in order to receive federal money for subsidized childcare (Seefeldt and Galper 1998). Before this legislation, the range of financial investment in early childhood education varied greatly from state to state. According to Adams and Sandfort in Seefeldt and Galper, state expenditures ranged from $0.24 per child in Idaho to more than $70 per child in Alaska,

**Table 1. Three- to Five-year-old Children Enrolled in Nursery School or Kindergarten (in percent)**

| Region | Three- to Five-Year-Olds Enrolled in Nursery School or Kindergarten (in percent) |
|---|---|
| **Pacific Northwest:** Alaska, Idaho, Oregon, Washington | 56% |
| **Southwest:** Arizona, California, Hawaii, Nevada, New Mexico, Texas | 58% |
| **Northwest:** Colorado, Kansas, Montana, Nebraska, North Dakota, Oklahoma, South Dakota, Utah, Wyoming | 56% |
| **Midwest:** Illinois, Indiana, Iowa, Michigan, Minnesota, Missouri, Ohio, Wisconsin | 60% |
| **Southeast:** Alabama, Arkansas, Florida, Georgia, Kentucky, Louisiana, Mississippi, North Carolina, South Carolina, Tennessee | 59% |
| **Northeast:** Delaware, District of Columbia, Maryland, New Jersey, New York, Pennsylvania, Virginia, West Virginia | 62% |
| **New England:** Connecticut, Maine, Massachusetts, New Hampshire, Rhode Island, Vermont | 66% |

Connecticut, Massachusetts, New York, and Vermont in fiscal year 1990 (Seefeldt and Galper 1998).

While particular regions have similar percentages of students participating in early childhood programs, many areas of the country have few or no programs available to their youth. One study reported by the Children's Defense Fund (1997) indicated that nine out of 55 counties in West Virginia had no childcare centers. Other studies have shown that childcare is particularly scarce in low-income communities. The U.S. Department of Education has found that public schools in low-income communities were less likely (16 percent) to offer preschool programs than their wealthier counterparts (approximately 33 percent). Similarly, only 33 percent of schools in low-income communities offered extended-day or enrichment programs, compared to 52 percent of wealthier schools (CDF 1997).

Not only do financial burdens prohibit children from participating in preschool programs, many students are denied such education because the waiting lists are simply too long. In 1995, 38 states and the District of Columbia had waiting lists of low-income working families who needed childcare assistance. In 1995 Texas had more than 35,000 children on its waiting list, constituting a wait as long as two years. Florida's waiting list recently reached almost 28,000, the highest it has been since 1991. Illinois had approximately 20,000 students waiting in 1995, many of whom will never come off the waiting list because priority is given to students needing protective services and those with special needs (CDF 1997).

### *Quality of Care*

According to Seefeldt and Galper (1998), most children are likely to be in satisfactory childcare situations. Even so, research indicates that the childcare offered to far too many children is, at best, inadequate and may even be harmful. All states fall short of providing high-quality childcare and education to all preschool students. In a study of 50 non-profit and 50 for-profit, randomly chosen childcare centers in California, Colorado, Connecticut, and North Carolina, researchers found that "only 1 in 7 centers provides a level of childcare quality that promotes healthy development and learning" (Cost, Quality, and Child Outcomes in Childcare Centers 1995). Consequently, 86 percent of the centers received ratings of poor or mediocre.

These findings are supported by data from *Who Cares? Childcare Teachers and the Quality of Care in America*, which indicates that many states are not truly committed to providing high-quality childcare and education for young children (Whitebook et al. 1989). The report's conclusions follow.

- Early education and childcare was a low priority for all states. Twenty-nine states spend less than 50 cents out of every $100 of state tax revenues on such programs. However, two-thirds of the states spend more than 10 times that amount on prisons and other correctional institutions.
- The commitment to implement high-quality early education programs varies greatly from state to state. The 10 states with the greatest commitment spent an average of 4.5 times as much money per child as the 10 states with the smallest commitment.
- Commitment by a state to early childhood education does not necessarily translate into high levels of spending on early childhood education. Kentucky, North Carolina, and Oklahoma were in the top third of states expressing commitment to high-quality early education in 1994, but they ranked in the bottom one-third of states in personal income per capita. Nevada and Virginia, on the other hand, were in the bottom one-third of states in terms of financial commitment, but they ranked 11th and 14th, respectively, in personal income per capita.

In order to achieve gains for students and consequently for society as a whole, states must commit more money to these programs. This means, among other things, that students from low-income families should have access to services that address not only educational issues, but social-service concerns as well. Because children's health has a tremendous impact on their development and readiness to learn, a complete education must also include health and nutrition services.

Research also shows that high-quality childcare is related to the staff-to-student ratio, the level of education of staff members, and teacher wages (Cost, Quality, and Child Outcomes in Childcare Centers 1995; Smith et al. 1995). The National Association for the Education of Young Children (NAEYC) recommends that an acceptable adult-to-child ratio for four- and five-year-olds is two adults with no more than 20 children (Bredekamp and Copple 1997). NAEYC also recommends in its *1993 Compensation Guidelines* that programs offer staff salaries and benefits commensurate with the skills and qualifications required for specific roles. Doing so would ensure the provision of high-quality services and the effective recruitment and retention of qualified, competent staff (Bredekamp and Copple 1997).

## Challenges for the Future: Providing A Safe and Caring Environment for Our Children

The importance of a quality preschool education must become a priority. Research has shown that the quality of early childhood education affects children's development and family relationships. The ultimate consequences of poor-quality childcare services are too great. If we do not pay now, we will pay later in the building of prisons and the loss of human capital (Seefeldt and Galper 1998). In order to rise above the mediocre childcare that is prevalent in most of America, we must insist on increased standards of quality.

Preschool children have different needs from older children, and these needs should be considered if preschool programs are going to be superior. Koralek, Colker, and Dodge discuss several key indicators of high-quality programs in their book, *The What, Why, and How of High-Quality Early Childhood Education: A Guide for On-Site Supervision* (1995). First, the program should be based on practice that is "developmentally appropriate." Although the primary role of education is the child's intellectual development, such growth cannot occur without fostering the child's social, emotional, and physical development. NAEYC identifies two dimensions of developmentally appropriate practice (Day 1994):

1. Age appropriateness. A predictable sequence of growth and development characterizes young children, and developmentally appropriate learning environments attend to what we know about how young children grow, develop, and learn.
2. Individual appropriateness. Developmentally appropriate practice affirms that each child is unique, with individual differences. Appropriate learning environments not only recognize the uniqueness of each child, but also reflect differences in the curriculum and experiential learning experiences offered to each child.

Developmentally appropriate learning environments are based on the following principles (Day 1994):

- Appropriate curriculum stimulates learning in all developmental areas: physical, social, emotional, and intellectual. Such a child-centered approach is at the heart of developmentally appropriate practice.
- Learning experiences are designed to support individual needs and differences. Developmental levels, learning styles, family backgrounds, and children's interests are among the factors that help formulate the learning environment. Differences among young children are evident. These differences should be acknowledged and used as a guide to inform instruction.

- Learning experiences provide children with the opportunity to actively manipulate and explore materials. Hands-on learning strategies emphasize the acquisition of higher-order critical thinking skills as opposed to drill-and-rote memorization. Young children learn by doing. They learn through exploration and discovery, using all of their senses. The optimum learning environment promotes active participation and provides many opportunities for children to see, feel, hear, smell, taste, and touch.
- Curriculum is designed to provide children with choices of many concrete and relevant learning experiences. If learning is relevant for children, they are more likely to persist with a task and are more motivated to learn.
- Learning opportunities are presented predominately in learning centers, where children work individually and in small groups, as opposed to whole-group instruction.
- Learning is viewed as integrated, and opportunities to develop math, science, and literacy skills can occur simultaneously rather than in discrete, segmented lessons. Units of study and topic work are used to present related and integrated curriculum (NAEYC as cited in Day 1994).

Another indicator of a high-quality preschool program is the degree to which the environment is safe and orderly. From making sure that play areas are free from hazardous materials to tending to the health and nutrition of the student, high-quality preschools concern themselves with issues of safety.

In their list of recommendations, Koralek, Colker, and Dodge (1995) also include the need for students to feel respected by their adult caretakers. In addition, parental involvement helps to facilitate quality in the preschool program.

Extensive research into how young children learn dictates that the following elements should be fundamental to a preschool program (Day 1994):

1. Throughout the preschool years, the curriculum should be presented in an integrated format rather than in 10- or 20-minute segments for each content area. Instruction should be planned around themes, with the themes being developed through learning centers in which the children are free to plan and select activities that support their individual learning experience.

2. Children in preschool should be engaged in active, rather than passive, learning activities. The curriculum must be seen as more than a program purchased from a publisher. This program should not dictate what learning is appropriate for a given

child. The teacher should serve as a facilitator who informs instruction for each individual.

3. Spontaneous play, either alone or with other children, is a natural way for young children to learn to deal with one another and to understand their environment. Play should be valued and included in the program plan.
4. Because children come to school with different knowledge, concepts, and experiences, it is important that new learning be connected to something that is known and relevant to them (National Association of Elementary School Principals as cited in Day 1994).

If we are to change the state of affairs for children in America, we must begin with high-quality, developmentally appropriate educational opportunities. While all students should reap the benefits of such programs, special care should be given to those who are financially burdened. If we are to take students into the new millenium equipped to be successful in school, changes will have to be made in the number of students who receive early childhood educational services and in the quality of the programs in which they participate. This goal will only become a reality through the implementation of programs that meet the criteria outlined in this paper.

## References and Bibliography

American Psychological Association. (1993). *Violence and youth: Psychology's response, Vol. 1: Summary report of the American Psychological Association Commission on Violence and Youth.* Washington, DC: Author.

Annie E. Casey Foundation. (1997). Kids Count: State profiles. Available online at: http://www.aecf.org/cig-bin

Bowman, B. (1994). The challenge of diversity. *Phi Delta Kappan*, 76(3).

Bredekamp, S., and Copple, C. (1997). *Developmentally appropriate practice in early childhood programs.* Washington, DC: National Association for the Education of Young Children.

Bridgman, A. (1989). *Early childhood education and childcare: Challenges and opportunities for America's public schools.* Arlington,VA: American Association of School Administrators.

Children's Defense Fund. (1997). *The state of America's children yearbook.* Washington DC: Author.

Cost, Quality, and Child Outcomes in Childcare Centers. (1995). *Executive summary.* Denver, CO: University of Colorado at Denver.

Day, B. (1994). *Early childhood education: Developmental/experiential teaching and learning (4th ed.).* New York: Macmillan College Publishing Company.

Koralek, D.G., Colker, L.J., and Dodge, D.T. (1995). *The what, why, and how of high-quality early childhood education: A guide for on-site supervision.* Washington, DC: National Association for the Education of Young Children.

National Education Goals Panel. (1993). *Building a nation of learners: The national education goals report: Executive summary.* Washington, DC: Author.

Seefeldt, C., and Galper, A. (1998). *Continuing issues in early childhood education.* Upper Saddle River, NJ: Prentice-Hall, Inc.

Sherman, A. (1994). *Wasting America's future: The Children's Defense Fund report on the costs of child poverty.* Boston, MA: Beacon Press Books.

Smith, S.L., Fairchild, M., and Groginsky, S. (1995). *Early childhood care and education: An investment that works.* Washington, DC: National Conference of State Legislatures.

Weikart, D.P. (1989). *Quality preschool programs: A long-term social investment.* New York: Ford Foundation Project on Social Welfare and the American Future, Occasional Paper Number Five.

Whitebook, M., Howes, C., and Phillips, D. (1989). *Who cares? Childcare teachers and the quality of care in America: Final report, National Childcare Staffing Study.* Oakland, CA: Childcare Employee Project.

Willer, B., and Bredekamp, S. (1990). Redefining readiness: An essential requisite for educational reform. *Young Children,* 5: 22–24.

# Policy Implications for Math, Science, and Technology In Early Childhood Education

*Barbara T. Bowman*

Math, science, and technology are not generally thought of as curricula for young children. Aside from counting, number recognition, growing plants, and learning food groups, math, science, and technology are generally given short shrift during the preschool years. Nevertheless, the roots of later competence are established long before school age, and recent findings from neuroscience confirm the importance of the link between early experience and subsequent achievement. Given the connection, why has so little attention been given to what and how young children acquire their knowledge of math and science or how teachers and parents can enhance it? The limited attention is not surprising because literacy commands so much time during the preschool and primary years that there is often little time left for other subjects. In addition, the knowledge base for math, science, and technology is formal, orderly, abstract, and based in a stern reality, whereas young children's learning is informal, relational, concrete, and often heavily imbued with fantasy. Nevertheless, how best to structure early learning to help children make connections to the formal disciplines is important and should be of interest to parents, educators, and policymakers.

Math, science, and technology are ubiquitous in our society. Even without organized curricula and instruction, children form an informal or intuitive base of knowledge of concepts in these disciplines through their interactions with people and things. They have many opportunities to consider numbers of objects, to observe physical phenomena, to use technologically driven equipment (telephones, televisions), and to make hypotheses regarding cause and effect. Although some of the concepts they form are

*Barbara T. Bowman is president of the Erikson Institute in Chicago.*

incorrect (presumably due to their immaturity and limited experience), the process of "doing" science and math and using technology is well established by the time children are three years old.

Fifty percent of U.S. children between the ages of three and five are enrolled in preschool programs that provide somewhat organized learning experiences. However, there are no state or nationally enforced preschool curricula, nor are there commonly accepted math, science, or technology standards in these programs. As with children who stay at home, the amount and quality of children's preschool experience is extremely dissimilar. Nevertheless, there are forces that exert pressure on preschools and parents and create some coherence in what young children are apt to learn. Some of the forces and trends that help frame early childhood curricula in math, science, and technology will be briefly reviewed in this paper.

The most contentious issues in education surround what to teach children and how to teach it. These issues have been bitterly contested throughout the century, primarily in public education, but increasingly in preschool education. Early childhood is defined as the years between three and eight, and educators have become concerned about discontinuities in the curricula between preschool, kindergarten, and the primary grades. At the core of the disagreements are different beliefs about development and learning and the needs of society.

## Theories of Development and Learning

Differences of opinion about how children learn and the content and methods for teaching have drawn adherents into bitter controversies, occasionally described as wars, and programs have made pendulum swings as one theory or another gained prominence. The "theory" disagreements have filtered into the preschool field as more children are enrolled in formal programs.

### *Constructivist Theory*

At present, the most influential theoretical position is often called "constructivist," or "progressive." It emphasizes the importance of young children constructing knowledge (understanding concepts) through their own activity, as opposed to simply being told correct answers by others. Children are encouraged to handle objects, observe and predict results, hear and use language, and collaborate with

adults and older children to develop ideas. Children's own motivations and concerns are critical in shaping the learning process since engagement of the learner and control of the pace and type of information to be processed are essential characteristics of this type of teaching and learning. Although children may temporarily make mistakes in understanding, it is important for adults to encourage children to learn through doing and to have confidence in their own ability to think. This perspective speaks to the importance of relationships with others that either confer on learning the mantle of adventure, discovery, meaningfulness, and pleasure or that of drudgery and monotony.

### *The Romantic View*

A second perspective stems from the "romantic" or "Rousseau" view of young children. According to this view, children are thought to learn best when following their own interests and inclinations. They are naturally curious about their world and will learn best through exploration, play, and practicing what they observe around them. Maturation, determined by the individual growth pattern of each child, is critical to understanding and enjoying formal learning. Forcing children to prematurely master adult knowledge systems before they are ready is thought to compromise their potential. Concern is often expressed about hurrying children into academic learning, rather than waiting until the child is mature enough to have laid a solid foundation. From this perspective, formal learning should be delayed until children demonstrate their cognitive, social, emotional, and physical readiness to learn in school, which will probably not occur until children are at least five or six years of age.

### *The Basic Skills Approach*

Another perspective, often called the "direct instruction" or "basic skills" approach, focuses on young children's ability to "learn by rote" without fully understanding the ideas being expressed. Children can learn to perform specific skills (such as counting and adding) without knowing the precise meaning of their activity. For instance, a young child may pronounce the number words correctly, but he or she cannot point to a set of objects whose quantity correlates with each number word. This inability indicates that he or she has not mastered the relationship between number words and the number of objects. Frequent practice of correct responses, however, is considered key to acquiring knowledge. Drill-and-practice sets the stage for understanding certain concepts, if it is not a precondition, for understanding. Therefore, learning to count by rote, for instance, lays the foundation for understanding numbers.

Advocates of this position identify a limited set of basic skills—counting, alphabet naming, number facts, etc.—that they believe young children can and should learn. They consider such learning to be the building blocks for solving problems.

### *The Social-Context Perspective*

A fourth perspective focuses on the social context in which math, science, and technology are embedded. The empirical nature of these disciplines has led many people to assume they are "culture free," which is certainly not the case. The basic assumptions about the nature of these disciplines are a part of Western culture and must be learned. Furthermore, children understand the personal meaning of math, science, and technology in the context of their own families and communities. Whether these disciplines are viewed as useful concepts for everyday life or as esoteric subjects only relegated to school comes through children's early interactions with their family, friends, and community members. Thus, children from different social classes and ethnic groups often do not learn about these disciplines in the same way as mainstream children, nor do they have the same expectations regarding their own acquisition of knowledge.

According to this view, many low-income and minority-group children have fewer opportunities to engage with knowledgeable adults about the disciplines, their parents and teachers do not set high expectations for their school learning, and their achievement suffers. This perspective argues for parent education and supports, so families can, in turn, support children's school achievement.

Support for all of these teaching/learning theories is easily found. Children explore and play with a variety of toys and materials, discuss with others how and why things work or happen, learn the conventions of the alphabet and numbers by rote in songs and games, and observe older members of their family use math, science, and technology concepts. What differs is the relative emphasis advocates of each perspective place on their own approach when structuring programs for children.

### *The Developmentally Appropriate Practices View*

Professional organizations have been among the most outspoken advocates of moving programs for young children toward more "progressive" theory. The National Association for the

Education of Young Children (NAEYC), under the rubric "developmentally appropriate practices" (DAP), has probably had the greatest influence on what and how preschool children are taught. DAP reflects "constructivist" theory and directs teachers' (and parents') attention to children's individual differences in temperament and the pace of growth, age-related abilities and learning styles, and cultural differences. Although no explicit curriculum is suggested, math, science, and technology are included as appropriate subject matter, with preference given for activities that are meaningful to children as opposed to having them learn primarily through drill and practice.

The National Association for the Education of Young Children, with more than 100,000 members, has actively disseminated this position to its members and to other professional organizations. The concept of developmentally appropriate practices has found broad acceptance in allied fields. It has been endorsed by the National Association of Early Childhood Specialists in state Departments of Education, the National Association of Elementary School Principals, and the Council for Exceptional Children's Early Childhood Division. Furthermore, it is consistent with the recommendations of state Departments of Education and a broad range of professional organizations, including the National Council of Teachers of Mathematics and the National Academy of Sciences.

The DAP perspective has been frequently misinterpreted. It is not a model but a set of principles on which to select curriculum based on a number of different factors, such as the age of the child, individual interests, and maturation rates. Unfortunately, and too often, the principles have been taken as laws to prohibit academic content in preschools. Therefore, math and science instruction has been left to the whims of children's interests rather than being cultivated by thoughtful and developmentally appropriate programming. NAEYC has protested this result as polarizing "into either/or choices many questions that are more fruitfully seen as both/and." The association has sponsored a number of initiatives designed to correct this impression among teachers, administrators, and the general public.

## Changes in Math, Science, and Technology

At least some of the heat about how and what to teach young children is generated because advocates have different beliefs about the future and the educational needs of citizens of the

21st century. The last quarter of the 20th century has seen a dramatic shift in how people organize their thoughts about the world, its resources, and their relationships with one another. The transformation in thinking is comparable to earlier paradigm shifts that took place during the Commercial and Industrial Revolution. Among the characteristics of this new revolution are:

- the reorganization of knowledge systems through the use of high-powered computers;
- increased access to inexpensive audiovisual and computer hardware and software, worldwide computer networks, and interactive technologies;
- new symbol systems (computer programming) and the use of old systems in new ways (word processing, calculators, etc.); and
- new models for understanding the world (artificial intelligence and problem simulations).

Many professionals in math, science, and technology envision new demands on their disciplines in the postindustrial world. They now assume a world in which new knowledge is quickly generated and even more quickly disseminated, thereby requiring students to have a broader range of skills and knowledge, to be more facile in their use of what they know, and to be ready to solve new problems. For instance, the National Council of Teachers of Mathematics (NCTM) published high standards for teaching math. NCTM criticized traditional curricula as being preoccupied with computation and rote activities and failing to foster mathematical insight, reasoning, and problem solving. A more serious problem, they contend, is that young children learn in the early grades that mathematics does not make much sense and they question their own ability to understand and learn.

The American Association for the Advancement of Science and the National Research Council of the National Academy of Sciences are similarly ambitious for the achievement of young children and also focus on problem solving, direct experience, and understanding as the primary goals of early childhood education. As with the NCTM standards, content ranges over an extensive list, including physical, biological, and chemical sciences, and encourages observation and simple experimentation. A separate standard focuses on helping students connect science and technology by giving them first-hand experience with technological products such as zippers, coat hooks, and can openers. Technology advocates are concerned that children learn to use computers as tools to extend their ability to solve problems rather than simply learning how to perform specific skills from pre-programmed problems and solutions. They endorse the idea that children should be engaged in meaningful activities in which they act as doers and thinkers rather than as recipients of information from teachers.

## Implementing Programs

Over the past quarter century, there has been increasing concern about the quality of education in the United States. Politicians and business leaders, in particular, have been concerned that American children are losing the competitive edge as compared to children from other industrialized countries, who often test better in math and science than American children. The National Governors Association, Congress, state governments, and local school districts have taken up the challenge to improve schools by setting high standards, assessing children's performance, and holding schools accountable. The result is that each state has devised its own standards, assessment strategies, and accountability measures. Local school districts are expected to demonstrate the steps they will take to achieve their state's goals. While these measures have resulted in new and creative practices in some schools, in others it has led to a call to return to traditional practices, which emphasize basic skills and whole-class, direct instruction, even in preschool.

Although state and local standards, assessment strategies, and accountability practices focus on kindergarten through 12th grade, they have influenced early childhood programs in a number of ways. The first National Education Goal defined by Congress and the nation's governors, "All children will come to school ready to learn," has energized public support for pre-kindergarten programs that serve at-risk children. As a consequence, national funding allocations for Head Start have risen, and states have increased their own funding for pre-kindergarten programs. In addition, the U.S. Department of Health and Human Services and various states and localities sponsor initiatives to influence the quality of daycare programs.

Pressure on public schools to improve programs has also resulted in more careful screening of children before they enroll in kindergarten, more strict age requirements, ability grouping, and a more stringent kindergarten curriculum. Parents are targeted for education programs (such as Parents As Teachers in Missouri) to encourage them to help prepare their children for school. Two-generation literacy programs as well as instruction in obtaining the Graduate Equivalency Diploma (GED) and work training have also been connected to early childhood centers. Unfortunately, these programs have proven only moderately effective. In some areas of the country, there has been pressure on parents and teachers to hold back children who are not socially mature and who are not achieving at the normative level—despite the lack of evidence that retention in preschool results in better performance in kindergarten.

Despite the overwhelming support of professional organizations for promoting curricula and methods that foster active and interactive learning and that motivate children to enjoy science and math activities, a backlash has built up in public education against the new standards and teaching methods. While there are probably many reasons for this reaction, at least one is the inadequacy of current assessment techniques. As the result of using standardized tests as the primary way of determining children's learning, many public schools (and parents and legislators) have advocated curricula and teaching methods that are more likely to raise children's scores. Whole-group, direct instruction is most in tune with this goal, given the financial constraints on many school districts. In addition, children's decreased ability to perform traditional computational skills has shaken the public's faith in new standards and methods.

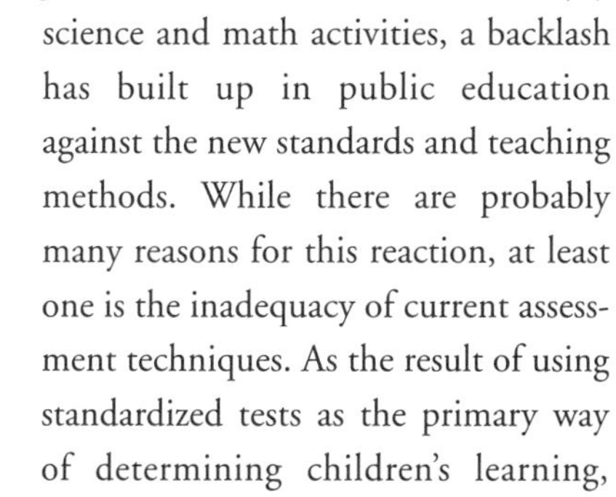

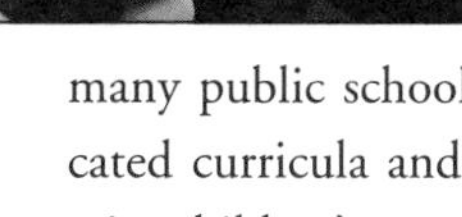

Another deterrent to implementing math and science standards is the inadequate preparation of teachers. Preparing children to meet the new high standards is a major challenge for teachers who, themselves, are unsure of the nature and implications of the change. Even college-level general education courses are often focused on operations, not problem solving. Preschool teachers, many of whom are paraprofessionals without a general education background, are even less able to implement the ambitious programs recommended by professional associations.

The media have taken up the banner of early learning and also present opportunities for young children to learn math, science, and technology. The groundbreaking television program, *Sesame Street*, is still the most widely viewed program by our young children. Although *Sesame Street* tries to make math and science knowledge meaningful to children in each program segment, the format still relies heavily on rote counting and alphabet naming. Other television programs, such as *Mr. Rogers*, fit math and science concepts into everyday contexts, focusing on meaning rather than memorized facts. Book publishers are similarly engaged in disseminating math and science ideas to young children. Again, many of these books stress rote counting and naming alphabets, but many also embed numbers and science concepts into stories and poems.

There is little evidence that any one television program, book, or computer program alone can account for what young children know and are able to do. However, it is clear that children who are exposed to multiple opportunities to learn through the media, books, and computers, as well as those who have knowledgeable family members and preschool teachers, are more likely to be well prepared for school than children who do not have such reinforcement.

## High-Risk Children

The first National Education Goal, "All children will start school ready to learn," implicitly recognizes the importance of good cognitive, physical, social, and emotional health if children are to achieve well in school. Some low-income and some minority-group children—primarily those who are African American, Latino, and Native American—arrive at school already disadvantaged since they are less likely to have the same opportunities to learn "academics" as do mainstream children. Children from low-income families are less likely to attend preschool than other children (by a gap of 24 percent to 52 percent). Head Start still does not serve all of the eligible low-income four-year-olds and also serves fewer than 20 percent of eligible three-year-olds.

Fortunately, the idea of high-quality early childhood education for children at-risk is gaining broad acceptance. However, children judged to be "at-risk" are more likely to participate in skill-centered programs than in ones that offer multiple, rich opportunities to "learn by doing." Driven by the wish to raise low scores on standardized tests, schools and learning centers frequently endorse drill-and-practice type activities. Many parents in low-achieving schools prefer such programs because they have clearer learning outcomes, outcomes that can be assessed by the parents as well as the school or center. Furthermore, programs that serve children who come from families and communities that do not use much math, science, and technology in their daily lives are less apt to focus on these disciplines. Nevertheless, most early childhood educators still support the notion of providing a greater variety of opportunities for children to learn through direct experience.

Young children share their community's perceptions of the place of math, science, and technology in the social world and the individual's relationship to them. There is probably nothing "inevitable" about the way these subjects are integrated into the social fabric of our society. Young children can learn which skills are "socially desirable" and expected of them or, conversely, what knowledge is exclusive and more available to some people than to others. In contemplating the social context of teaching, policymakers must be mindful that institutions tend to duplicate current power relationships among people. They must consider the effect, for example, of offering middle-class children opportunities to play with ideas while providing few such chances for poor children. What seems to be needed is a balance between skills and intellectual content and between teacher-directed curricula and children's own interests and motivation to learn. Children may learn to view math and science concepts as "the single truth" available to a few or as one way of looking at the world that is open to all. The choice is ours to offer.

# Concept Development in Preschool Children

*Susan A. Gelman*

This paper addresses concept development in preschool children, based on recent psychological research. Over the past 30 years, there have been more than 7,000 journal articles written on children's concepts or categories. Scholars are attracted by the opportunity to understand fundamental theoretical issues (How can we characterize early thought? How does it change over time?) as well as by the practical concern of determining how children reason about concepts that are directly relevant to their lives and schooling (including mathematics, biology, and physics).

Four key themes have emerged from recent research. They will be highlighted and illustrated in this paper.

- Theme 1. Concepts are tools and as such have powerful implications for children's reasoning—both positive and negative.
- Theme 2. Children's early concepts are not necessarily concrete or perceptually based. Even preschool children are capable of reasoning about non-obvious, subtle, and abstract concepts.
- Theme 3. Children's concepts are not uniform across content areas, across individuals, or across tasks.
- Theme 4. Children's concepts reflect their emerging "theories" about the world. To the extent that children's theories are inaccurate, their conceptions are also biased.

These four themes contradict some widely held (but erroneous) views of early concepts, and they raise a variety of issues regarding early education.

## Theme 1: Concepts as Tools

Concepts provide an efficient way of organizing experience. If children were *unable* to categorize, their experiences would be chaotic—filled with objects, properties, sensations, and events

*Susan A. Gelman is a professor of psychology at the University of Michigan–Ann Arbor.*

too numerous to hold in memory. In contrast to this hypothesized "blooming, buzzing confusion" (to use the words of William James), children from earliest infancy form categories that are remarkably similar to those of adults. Before they have even begun to speak, infants form categories of faces, speech sounds, emotional expressions, colors, objects, animals, and mappings across modalities (see Gelman 1996 for review). By 18 months of age, most children have begun a vocabulary "explosion," adding roughly nine new words each day to their vocabulary (Carey 1978). Assuming that most new words encode concepts, this fact suggests that one- and two-year-old children are adept at concept acquisition.

However, concepts do more than organize information efficiently in memory. They also serve an important function for a range of cognitive tasks, including identifying objects in the world, forming analogies, making inferences that extend knowledge beyond what is already known, and conveying core elements of a theory. Many of these tasks are central to school performance; thus, concepts can be thought of as the building blocks to these more complex skills.

One of these cognitive functions, known as *induction*, is the focus of the following discussion. Induction involves how concepts foster inferences about the unknown.

Both children and adults use categories to extend knowledge beyond what is obvious or already known (Carey 1985; Gelman and Markman 1986, 1987). For example, if four-year-old children are told a new fact, such as that a particular dog has leukocytes inside it, they are likely to infer that other dogs also have leukocytes inside them. It is important to note that children form such inferences even when they are not supported by outward appearances. In one Gelman and Markman example items children saw a brontosaurus, a rhinoceros, and a triceratops, which were labeled as "dinosaur," "rhinoceros," and "dinosaur," respectively. Category labels and outward appearances conflicted: The brontosaurus and triceratops are members of the same category, whereas the rhinoceros and triceratops look more alike outwardly. Then children learned a new property of the brontosaurus and the rhinoceros—that they had cold blood and warm blood, respectively. They were asked which property was true of the triceratops. When children were informed that both the brontosaurus and the triceratops were dinosaurs, they inferred that the triceratops has cold blood like the brontosaurus, even though it more closely resembled the rhinoceros. The results of this and other related experiments showed that by 2-1/2 years of age, children base inferences on category membership, despite conflicting surface appearances (Gelman and Coley 1990).

Although induction can be viewed as positive because it allows children to expand their knowledge base, it also poses some problems for young children when they draw inappropriately broad inferences. One problem that results is stereotyping. Preschoolers often treat social categories as if they were biological categories, assuming, for example, that members of a social category (a category that is based on gender or race) will be alike with respect to ability or occupation (Hirschfeld 1996; Taylor 1996). A second problem is that young children at times ignore relevant information about statistical variation *within* a category (Lopez et al. 1992; Gutheil and Gelman 1997). For example, four-year-olds do not seem to realize that a property known to be true of five birds provides a firmer basis of induction than a property known to be true of only one bird. They also do not seem to realize that *variability* in a category is relevant to the kinds of inductions that are plausible.

In sum, this theme illustrates four important points:

- Concepts are used by children and adults to extend known information to previously unknown cases through a process called *inductive inference.*
- Such inferences are not based on perceptual similarity alone.
- Naming is an important vehicle for conveying category membership and thus guiding induction. Naming leads children to search for similarities among category members. Thus, this function is a highly useful tool available to children by at least preschool age.
- Despite children's ability to use categories for induction, even in the preschool years, they do not always appropriately constrain these inferences.

## Theme 2: Non-Obvious Concepts

On many traditional accounts, conceptions are said to undergo a fundamental, qualitative shift with development. That is, children and adults are often said to occupy opposite endpoints of various dichotomies, moving from perceptual to conceptual (Bruner et al. 1966), from concrete to abstract (Piaget 1951), or from similarity to theories (Quine 1977).

These developmental dichotomies are intuitively appealing, in part because children often do seem to reason in ways that are strikingly different from how adults reason. For example, in the well-known "conservation error" studied by Piaget, children under six or seven years of age report that an irrelevant transformation leads to a change in quantity (e.g., concluding that the amount of a liquid increases when it is simply poured from a wide container into a taller, narrower container). Children appear to focus on one salient but misleading dimension—for example, the height of a container—forgoing a deeper conceptual analysis. Throughout the past several decades, there have been many demonstrations that young children are "prone to accept things as they seem to be, in terms of their outer, perceptual, phenomenal, on-the-surface characteristics" (Flavell 1977).

However, as an account of what children are *capable* of doing, such developmental dichotomies as the "perceptual-to-conceptual shift" are inadequate (also Bauer and Mandler 1989; Gelman 1978; Gibson and Spelke 1983; Markman and Hutchinson,1984; Nelson 1977). With appropriately sensitive tasks, children can display abilities that do not appear in their everyday actions. (See Theme 3, Expertise and Task Effects Across Areas. See also Gelman and Baillargeon 1983 for extensive discussion.)

Indeed, not only are children capable of overcoming misleading appearances, they are even able to reason about concepts that are altogether non-obvious. Several scholars have recently begun studying how three- to five-year-old children reason about non-obvious entities across a range of topic areas, including:

- energy (Shultz 1982);
- bodily organs such as muscles, bones, and brains (Gelman 1990; Johnson 1990);
- germs and contagion (Kalish 1996);
- inheritance (Springer 1992);
- dissolved substances and contamination (Au 1994); and
- mental entities such as thoughts and desires (Wellman 1990).

Although for most of these topics preschool children have very little in the way of detailed, concrete knowledge, they have begun to appreciate *that* these non-obvious constructs exist and *how* they affect other, more observable outcomes and behaviors. For example, even three-year-olds have a core understanding that "germs" can cause illness and that foods may appear clean, yet still have disease-causing germs (Kalish 1996). It is intriguing that children are capable of this understanding at an age when they have not yet learned anything about the mechanisms by which viruses and bacteria affect human physiology (Au and Romo 1998). That children are open to reasoning about these topics and that they do so with considerable accuracy, despite an impoverished knowledge base, argues strongly that non-obvious entities are *not* beyond the capacity of preschool children.

## Theme 3: Expertise and Task Effects Across Areas

The previous section illustrated that detailed knowledge is not a prerequisite for learning some of the core concepts in a domain (such as "germs"). Nonetheless, specialized knowledge can exert surprisingly powerful effects on cognition (Hirschfeld and Gelman 1994; Wellman and Gelman 1997). Twenty-five years ago, Chase and Simon (1973) found that chess experts have superior memory for the position of pieces on a chessboard, although they are no better than non-experts in their memory for digits. Chi (1978) demonstrated the same phenomenon in children: Child chess experts even outperform adult chess novices, which is an interesting reversal of the more usual developmental finding. In these examples, experts are not *in general* more intelligent or more skilled than novices. The effects are localized within the domain of expertise.

Regarding concept development, too, the child's level of sophistication varies markedly by content area (Keil 1989, in press). Chi, Feltovich, and Glaser (1981), focusing on the domain of physics, found that expert problem-solvers and novice problem-solvers approach word problems about physics quite differently. They focus on altogether different features of the task. Similarly, Chi, Hutchinson, and Robin (1989) found that dinosaur experts who are children perform differently from novices in how they reason about the domain of dinosaurs; they generate a richer set of inferences and causal explanations.

Conceptual structure varies not only by domain but also by task. Even within a domain, children use different information in different contexts, depending on the task or function at hand (Deak and Bauer 1995, 1996; Taylor and Gelman 1993). For example, children display markedly different strategies for sorting pictures into groups, depending on whether the researcher provides the standard open-ended instructions ("See this picture? Can you find another one?") or whether the researcher teaches the child a new word and asks the child to extend it ("See this dax? Can you find another dax?") (Markman 1989). Similarly, children provide altogether different responses when they are asked to come up with the category label for an ambiguous item versus when they are asked to come up with a property inference for an ambiguous item that has been labeled by the experimenter (Gelman et al. 1986). Children consider a variety of information and flexibly deploy different sorting strategies depending on the task; they display links between task and strategy that are precise and predictable. These variations in performance are not random or idiosyncratic. Rather, they reveal that concepts have multiple functions, even for preschool children, and that children have learned to attend to different kinds of information, depending on the task at hand.

## Theme 4: Concepts and Theories

Adults' concepts are influenced by theoretical belief systems. This statement can be readily illustrated with a very simple example. Until recently, a biological mother could be defined or understood as the woman who gives birth to a child. More recently, however, with new reproductive technology (including surrogate mothers and donor eggs), a biological mother need not be the woman who gives birth. Thus, even a concept so basic and fundamental as "mother" undergoes change as one's theory of reproduction changes (which itself is influenced by changing technology).

A central developmental question is when and how children begin to incorporate theories into their concepts. One long-held view was that children's initial categories are similarity-based and that children only begin to incorporate theories as they gain experience and formal schooling (Quine 1977). Likewise, Piaget argued that pre-operational children do not have the logical capacity to construct either theories or true concepts.

In contrast, many researchers now believe that concept acquisition in childhood may *require* theories. Murphy (1993) notes that theories help concept learners in three respects:

- Theories help identify those features that are relevant to a concept.
- Theories constrain how (e.g., along which dimensions) similarity should be computed.
- Theories can influence how concepts are stored in memory.

The implication here is that concept acquisition may proceed more smoothly with the help of theories. If true, it is reasonable to expect that theories may play a role in children's concepts, even though the theories themselves are changing developmentally.

Indeed, recent studies provide compelling demonstrations that young children use theoretical knowledge to decide how to categorize data. Barrett, Abdi, Murphy, and Gallagher (1993) presented data suggesting that children's intuitive theories help determine which properties and which feature correlations children select in making classifications. For example, in a task that required children to categorize novel birds into one of two novel categories, first- and fourth-grade children noticed the association between brain size and memory capacity and used that correlation to categorize new members. Specifically, exemplars that preserved the correlation were more often judged to be category members and to be more typical of the category. The children did not make use of features that correlated equally well—features that were presented together equally often in the input—but that were not supported by a theory—for example, the correlation between the structure of a heart and the shape of a beak.

In a second experiment, Barrett et al. (1993) found that children selected different feature-pairs, depending again on whether they were supported by a pre-existing theory. Third-grade children were presented with hypothetical categories that were described as either animals or tools. They then learned five properties about each category. When the category was described as an animal, children selectively focused on correlations between one subset of the properties, for example, "is found in the mountains" and "has thick wool." In contrast, when the category was described as a tool, children selectively focused on correlations between a different subset of properties, for example, "is found in the mountains" and "can crush rocks." These studies show that children focus their attention strategically on information that is relevant to the implicit theory they have formed.

More generally, errors in children's theories may constrain or shape children's concepts. One brief example illustrates children's number concepts. Rochel Gelman has found that for preschool children and early elementary school children their understanding of arithmetic is heavily influenced by the theory that numbers are *countable* entities, starting with one and continuing sequentially by adding whole numbers. Based on this assumption, children experience tremendous difficulty when first encountering fractions, often treating them as if they were whole numbers (Gelman and Williams 1997). For example, when asked to sort cards (representing different amounts) onto a number line, children will treat a card with 1-1/2 circles on it as if it represented "2" on the number line. The difficulty children have here is not simply that they are encountering a new set of mathematical operations but that their prior theory clashes with the new system they have encountered. Similarly, theory-driven errors can be found in young children's reasoning about physics (Kaiser et al. 1986) and biology (Carey 1985; Coley 1993).

Although a number of studies have begun to examine the influence of theories on early concepts, little work addresses the reverse question, that is, the influence of concepts on early theories. However, it seems plausible that certain conceptual assumptions may constrain aspects of children's theories. For example, children appear to hold an "essentialist" assumption about categories (Gelman et al. 1994; Medin and Ortony 1989; Atran 1993), treating members of a category as if they have an underlying "essence" that can never be altered or removed. Essentialism is more compatible with creationist views than with evolutionary views of species origins (Mayr 1988), and it may even discourage children's learning of evolutionary theory (Evans 1994). Thus, the structure of early concepts may have broader implications for science education.

## Conclusions

The brief summary in this paper shatters several myths about children's early concepts, including

- Myth 1. The sole function of concepts is to organize experience efficiently.
- Myth 2. There is qualitative change in children's concepts over time, with major shifts between four and seven years.
- Myth 3. Until about age 7, most children are unable to reason about abstract concepts or non-obvious features.
- Myth 4. Children's concepts start out perceptually-based, becoming conceptual with development.

To the contrary, recent evidence documents that even preschool children make use of concepts to expand knowledge via inductive inferences, that children's concepts are heterogeneous and do not undergo qualitative shifts during development, and that children's concepts incorporate non-perceptual elements from a young age.

Given that children's concepts are in fact far more sophisticated than has been traditionally assumed, it becomes all the more important to ensure that early education exploits the capabilities that young children have. At the same time, given the close link between early concepts and emerging theories, one of the central challenges is to help children overcome pervasive faulty theories, some of which appear to persist into adulthood.

## References and Bibliography

Atran, S. (1993). *Cognitive foundations of natural history: Towards an anthropology of science.* Cambridge, MA: Cambridge University Press.

Au, T.K. (1994). Developing an intuitive understanding of substance kinds. *Cognitive Psychology*, 27:71–111.

Au, T.K., and Romo, L.R. (1996). Building a coherent conception of HIV transmission: A new approach to AIDS education. In *The psychology of learning and motivation: Advances in research and theory*, ed. D.L. Medin, 193–241. New York: Academic Press.

Au, T.K., and Romo, L.R. (1998). In *The psychology of learning and motivation: Advances in research and theory*, ed. D.L. Medin, 193–241. New York: Academic Press.

Barrett, S.E., Abdi, H., Murphy, G.L., and Gallagher, J.M. (1993). Theory-based correlations and their role in children's concepts. *Child Development*, 64:1595–1616.

Bauer, P.J., and Mandler, J.M. (1989). Taxonomies and triads: Conceptual organization in one- to two-year-olds. *Cognitive Psychology*, 21:156–184.

Bruner, J.S., Olver, R.R., and Greenfield, P.M. (1966). *Studies in cognitive growth*. New York: Wiley.

Carey, S. (1978). The child as word learner. In *Linguistic theory and psychological reality*, eds. J. Bresnan, G. Miller, and M. Halle, 264–293. Cambridge, MA: MIT Press.

Carey, S. (1985). *Conceptual change in childhood*. Cambridge, MA: MIT Press.

Chase, W.G., and Simon, H.A. (1973). Perception in chess. *Cognitive Psychology*, 4: 55–81.

Chi, M.T.H. (1978). Knowledge structure and memory development. In *Children's thinking: What develops?* ed. R. Siegler, 73–96. Hillsdale, NJ: Erlbaum.

Chi, M.T.H., Feltovich, H.A., and Glaser, R. (1981). Categorization and representation of physics problems by experts and novices. *Cognitive Science*, 5: 121–152.

Chi, M.T.H., Hutchinson, J., and Robin, A. (1989). How inferences about novel domain-related concepts can be constrained by structured knowledge. *Merrill-Palmer Quarterly*, 35: 27–62.

Coley, J.D. (1993). *Emerging differentiation of folkbiology and folkpsychology: Similarity judgments and property attributions*. Doctoral dissertation, University of Michigan, Ann Arbor.

Deak, G.O., and Bauer, P.J. (1995). The effects of task comprehension on preschoolers' and adults' categorization choices. *Journal of Experimental Child Psychology*, 60: 393–427.

Deak, G.O., and Bauer, P.J. (1996). The dynamics of preschoolers' categorization choices. *Child Development*, 67:740–767.

Evans, E.M. (1994). *God or Darwin? The development of beliefs about the origin of species.* Doctoral dissertation, University of Michigan, Ann Arbor.

Flavell, J. (1977). *Cognitive development.* Englewood Cliffs, NJ: Prentice-Hall.

Gelman, R. (1978). Cognitive development. *Annual Review of Psychology*, 29: 297–332.

Gelman, R. (1990). First principles organize attention to and learning about relevant data: Number and the animate-inanimate distinction as examples. *Cognitive Science*, 14: 79–106.

Gelman, R., and Baillargeon, R. (1983). A review of some Piagetian concepts. In *Handbook of child psychology, Vol. 3*, eds. J.H. Flavell and E.M. Markman, 167–230. New York: Wiley.

Gelman, R., and Williams, E.M. (1997). Enabling constraints for cognitive development and learning: Domain specificity and epigenesis. In *Handbook of child psychology, 5th ed., Cognitive development*, eds. D. Kuhn and R. Siegler, 575–630. New York: Wiley.

Gelman, S.A. (1996). Concepts and theories. In *Perceptual and cognitive development*, eds. R. Gelman and T.K. Au. New York: Academic Press.

Gelman, S.A., and Coley, J.D. (1990). The importance of knowing a dodo is a bird: Categories and inferences in 2-year-old children. *Developmental Psychology*, 26: 796–804.

Gelman, S.A., Coley, J.D., and Gottfried, G.M. (1994). Essentialist beliefs in children: The acquisition of concepts and theories. In *Mapping the mind: Domain specificity in cognition and culture*, eds. L.A. Hirschfeld and A.S. Gelman, 341–365. Cambridge, MA: Cambridge University Press.

Gelman, S.A., Collman, P., and Maccoby, E.E. (1986). Inferring properties from categories versus inferring categories from properties: The case of gender. *Child Development*, 57: 396–404.

Gelman, S.A., and Markman, E.M. (1986). Categories and induction in young children. *Cognition*, 23: 183–209.

Gelman, S.A., and Markman, E.M. (1987). Young children's inductions from natural kinds: The role of categories and appearances. *Child Development*: 58, 1532–1541.

Gibson, E.J., and Spelke, E.S. (1983). The development of perception. In *Handbook of child psychology, 4th ed., Vol. 3, Cognitive development*, eds. J.H. Flavell and E.M. Markman, 1–76. New York: Wiley.

Gutheil, G., and Gelman, S.A. (1997). Children's use of sample size and diversity information within basic-level categories. *Journal of Experimental Child Psychology*, 64: 159–174.

Hirschfeld, L.A. (1996). *Race in the making: Cognition, culture, and the child's construction of human kinds*. Cambridge, MA: MIT Press.

Hirschfeld, L.A., and Gelman, S.A. (1994). Toward a topography of mind: An introduction to domain specificity. In *Mapping the mind: Domain specificity in cognition and culture*, eds. L.A. Hirschfeld and S.A. Gelman, 3–35. Cambridge, MA: Cambridge University Press.

Johnson, C.N. (1990). If you had my brain, where would I be? Children's understanding of the brain and identity. *Child Development*, 61: 962–972.

Kaiser, M.K., McCloskey, M., and Proffitt, D.R. (1986). Development of intuitive theories of motion: Curvilinear motion in the absence of external forces. *Developmental Psychology*, 22: 1–5.

Kalish, C.W. (1996). Preschooler's understanding of germs as invisible mechanism. *Cognitive Development*, 11: 83–106.

Keil, F.C. (1989). *Concepts, kinds, and cognitive development*. Cambridge, MA: MIT Press.

Keil, F.C. (in press). In *Conceptual development: Piaget's legacy*, eds. E. Scholnick, K. Nelson, P. Miller, and S.A. Gelman. Mahwah, NJ: Erlbaum.

Kuhn, D., and Siegler, R. eds. *Handbook of child psychology, 5th ed., Cognitive development*. New York: Wiley.

Lopez, A., Gelman, S.A., Gutheil, G., and Smith, E.E. (1992). The development of category-based induction. *Child Development*, 63: 1070–1090.

Markman, E.M. (1989). *Categorization and naming in children: Problems of induction*. Cambridge, MA: MIT Press.

Markman, E.M., and Hutchinson, J.E. (1984). Children's sensitivity to constraints on word meaning: Taxonomic versus thematic relations. *Cognitive Psychology*, 16: 1–27.

Mayr, E. (1988). *Toward a new philosophy of biology: Observations of an evolutionist*. Cambridge, MA: Belknap Press of Harvard University Press.

Medin, D., and Ortony, A. (1989). Psychological essentialism. In *Similarity and analogical reasoning*, eds. S. Voxniadou and A. Ortony, 179–195. Cambridge, MA: Cambridge University Press.

Murphy, G.L. (1993). Theories and concept formation. In *Categories and concepts: Theoretical views and inductive data analysis.* New York: Academic Press.

Nelson, K. (1977). The syntagmatic-paradigmatic shift revisited: A review of research and theory. *Psychological Bulletin,* 84: 93–116.

Piaget, J. (1951). *Play, dreams, and imitation in childhood.* New York: Norton.

Quine, W.V. (1977). Natural kinds. In *Naming, necessity, and natural kinds,* ed. S.P. Schwartz, 155–175. Ithaca, NY: Cornell University Press.

Shultz, T.R. (1982). Rules of causal attribution. *Monographs of the Society for Research in Child Development,* 47: 1–51.

Springer, K. (1992). Children's awareness of the biological implications of kinship. *Child Development,* 63: 950–959.

Taylor, M.G. (1996). The development of children's beliefs about social and biological aspects of gender differences. *Child Development,* 67: 1555–1571.

Taylor, M.G., and Gelman, S.A. (1993). Children's gender- and age-based categorization in similarity judgments and induction tasks. *Social Development,* 2: 104–121.

Wellman, H.M. (1990). *The child's theory of mind.* Cambridge, MA: MIT Press.

Wellman, H.M., and Gelman, S.A. (1997). Knowledge acquisition. In *Handbook of child psychology, 5th ed., Cognitive development,* eds. D. Kuhn and R. Siegler, 523–573. New York: Wiley.

# Educating Young Children in Math, Science, and Technology

*David Elkind*

Any intellectually responsible program to instruct young children in math, science, and technology must overcome at least three seemingly insurmountable obstacles. One of these is our adult inability to discover, either by reflection or analysis, the means by which children acquire science and technology concepts. Another obstacle is that young children think differently than we do and do not organize their world along the same lines as do older children and adults. Finally, young children have their own curriculum priorities and construct their own math, science, and technology concepts. These concepts, while age-appropriate, may appear wrong from an adult perspective. We need to consider each of these obstacles before turning to a few suggestions as to how they can be best overcome.

*David Elkind is a professor in the Department of Child Development at Tufts University in Medford, MA.*

## Obstacle 1. Using Reflective or Logical Analysis

Math, science, and technology are abstract mental constructions far removed from the immediate, here-and-now world of the young child. As adults, we cannot retrace the steps we took in attaining these concepts inasmuch as they are part of our intellectual unconscious and unavailable to retrospective analysis. A simple thought experiment illustrates the point.

Imagine you are teaching a young child of four or five to ride a small, two-wheeled bicycle. What is the most important thing the child has to learn to succeed at this skill? If the reader is like most adults, he or she will answer "Balance. The child has to learn to keep his or her weight centered on the seat."

In fact, if you actually attempt to teach a child to ride a two-wheeled bicycle, the problem turns out to be quite different. What you observe is that the young child focuses either upon pumping the bicycle's pedals and forgets to steer, or focuses upon steering and forgets to pump. In fact, balance is attained when children coordinate pumping with steering. Once we have attained that balance, however, we are no longer aware of how we accomplished it. Although this example is a simple illustration, it makes a very powerful point. If we want to teach young children math, science, and technology, we cannot start from some reflective analysis of the task, but rather we must actually observe children attempting to learn the task. This statement reflects one of Jean Piaget's (1950) most important insights, and one that must not be forgotten.

To make this insight concrete, consider Piaget's investigations in his book entitled *The Child's Conception of Number* (1942). In that text, he nicely demonstrated the parallel between the child's spontaneous construction of number and the three types of scales used in psychological measurement. In psychological research, we distinguish between nominal, ordinal, and interval scales of measurement. We speak of a *nominal* number when we use number as a name, such as the number on a football or basketball jersey. Nominal numbers have no numerical value or meaning. Secondly, we speak of an *ordinal* number when we use a number to designate a rank or to establish an order. The numbers used to describe a figure skater's performance are ranks. That is to say, a rating of 5.6 given one skater is better than a rating of 5.4 given to a second, but there is no exact measure of how much better the one skater is than the other. There are no units of skating skill or of artistic presentation.

Finally, we speak of an *interval* number when the numbers we use reflect equal units or intervals. It is only interval numbers that justify the operations of arithmetic and higher mathematics. This fact is often neglected, however. Many psychological measurements—such as the IQ, which is number used as a rank—are treated as if they were interval measures. The distinction between nominal, ordinal, and interval scales dates from the early days of recognizing psychology as a science and its attempts to employ quantitative methods. Interestingly enough, as Piaget has shown, children employ first nominal, then ordinal, and finally interval scales in their spontaneous attainment of measurement concepts.

Inasmuch as a nominal number is essentially a label, young children can use nominal numbers as soon as they are able to use names, usually by age two or three. Then, by the age of three or four, young children are able to order blocks (or other size-graded materials) as to size, and thus they demonstrate the ability to construct ordinal scales and employ numbers in an ordinal sense. It is, however, only when children attain what Piaget termed *concrete operations*, at ages five to seven, that children can construct units and employ interval scales.

The way in which a child constructs a unit illustrates how different this process is than what one might conclude on the basis of an introspective or logical analysis of the task. Piaget (1942) demonstrated that to construct the concept of a unit the child must coordinate the ideas of sameness and difference. To understand a unit the child must grasp the idea that, for example, the number three is both like every other number used in counting (its cardinal meaning) but also different from every other number (its ordinal meaning), because it is the only number that comes after two and before four. It is only at this stage that the child can perform true arithmetic operations. In short, the only way to understand how children learn a concept is to observe them in the process of acquiring it.

### *Obstacle 2. Young Children's Transductive Thinking*

Young children think differently than do older children and adults. It was Barbel Inhelder and Jean Piaget (1958) who gave us our most important insights into the thinking of young children. What Inhelder and Piaget discovered, among other things, was that young children think *transductively*, from object to object and from event to event, rather than inductively (using a set of specific facts to reach a general conclusion) or deductively (applying general principles to specific facts to reach a conclusion). All concepts and ideas are at the same level. For example, a child who asks, "If I eat spaghetti, will I become Italian?" is thinking transductively, joining concepts at two very different levels of abstraction. Transductive thinking, it has to be emphasized, is age-appropriate and is not something to be overcome or extinguished.

Transductive thinking helps to account for a number of characteristics of the thought process of preschool children. Young children often exhibit *animism* and ascribe life to any object that moves. Again, this thinking arises from the joining of concrete and abstract conceptions as if they were on the same level. Young children also give evidence of *purposivism*, the idea that every event and object has a purpose or goal. Finally, young children also give evidence of *phenomenalistic causality*, the idea that when two things occur together, one causes the other.

In many ways, Inhelder and Piaget have done for child thought what Kuhn said historians of science have begun doing for science.

> Rather than seeking the permanent contributions of an older science to our present vantage, they attempted to display the historical integrity of that science in its own time...Furthermore, they insist upon studying the opinions of that group and similar ones from the viewpoint—very different from that of modern science—that gives those opinions the maximum internal coherence and the closest possible fit to nature. (Kuhn 1962/1970, page 3)

In the same way, young children's thinking has to be understood on its own terms and in its own context, not from the perspective of adult thought.

### Obstacle 3. The Fundamental Curriculum

The third obstacle to the effective math, science, and technology education of young children is that preschoolers have their own curriculum goals. As adults we tend to assume that young children are born with all of the concepts that most children display when they enter first grade. We reach this conclusion because as adults we do not have many memories of the first few years of our lives. Recall memory requires a space-time framework that young children have yet to achieve. It is not until the age of seven or eight that children have a good sense of clock time. A true understanding of calendar time comes even later than that. In the same way, young children only acquire a sense of map and geographical space when they are in the later elementary grades. Without a space/time conception, there is no framework within which to order and store memories.

What we learn as young children, despite not remembering it, is what might be called the *fundamental curriculum* (Elkind 1987): our knowledge of things, their sensory properties, their spatial relations, and their temporal sequencing. Put more concretely, to operate successfully in the world, young children must learn the concepts of *light* and *heavy*, *behind*, *inside* and *on top*, *night* and *day*, *before* and *after*, and much, much more. None of these ideas is inborn; they all must be constructed using a great deal of time and effort. Young children have their own curriculum priorities.

Perhaps Friedrich Froebel (1904), the inventor of the kindergarten, put it best when he wrote that young children need to "learn the language of forms before they learn the language of words." Even without explicit instruction, young children are acquiring elementary and adaptive knowledge and skills in math, science, and technology.

## Implications for Math, Science, And Technological Education

The foregoing obstacles to math, science, and technological education have a number of practical implications for educating children in these domains. Only a few of these implications can be described in this paper. They are the importance of observing young children learning, the need to recognize the limits of instruction, and the value of employing capacity-linked and socially derived motivation.

### *Observing Young Children Learning*

In a nursery school recently, I observed a group of children gathered around a computer that one child was operating. They were working together and making suggestions to the child at the keyboard, who seemed to appreciate their help. The teacher, however, intervened and suggested that they needed to take turns and to let the child at the keyboard have his turn without the other children bothering him. This example shows how a well-intentioned teacher nonetheless ignored the necessity of observing children before making an instructional decision. The children were working cooperatively, not fighting or competing to be at the keyboard. Young children may approach technology in this way; they may find it less intimidating when approached as a group project. That at least is a possibility that should be investigated.

The world of technology, particularly computers, is a new one. But the obstacle to the most effective instruction using this technology is the same. We need to observe how children themselves deal with the technology. To be sure, some initial instruction is required and certain limits need to be set, but we also need to be careful observers of the choices that children themselves make. To give a second illustration, in another nursery school, a child at a computer was enjoying an animated reading program that she chose. The teacher, however, encouraged her to use a more advanced, strictly word-oriented program. In such situations, we need to respect children's choices. We really don't know what types of programs are most effective with young children without careful observation and study.

### *Recognizing the Limits of Young Children's Learning*

Transductive thinking is concrete and unilevel. Math, science, and technology, on the other hand, have concepts at many different levels of abstraction. For some disciplines, even the lowest levels of abstraction are beyond the abilities of young children. The failure to recognize and accept this fact was the fundamental error in Jerome Bruner's (1962) assertion that, "you can teach any child any subject at any age in an intellectually responsible way." There is simply no way to teach a preschool child algebra without so concretizing the concepts to the point that they appear non-algebraic. Algebra requires the learner to have acquired what Inhelder and Piaget (1958) call *formal operations*, operations that enable the young person to deal with second symbol systems. In algebra, a letter stands for another symbol, a number. Young children are unable to deal with second symbol systems.

A more concrete example may help amplify the point and provide a practical guideline. Early childhood is a question-asking period, but how one answers these questions has to reflect the child's level of thinking. If a child asks, "Why does the sun shine?" he or she will be lost if you begin to explain the relation between heat and light. The young child is not asking for a scientific explanation but for a purposive one. If we say, "To keep us warm and make the grass grow," we respond to the true intent of the child's question. Such answers are not really "wrong," and they accomplish the important goal of encouraging further questioning while providing a wonderful sense of being understood.

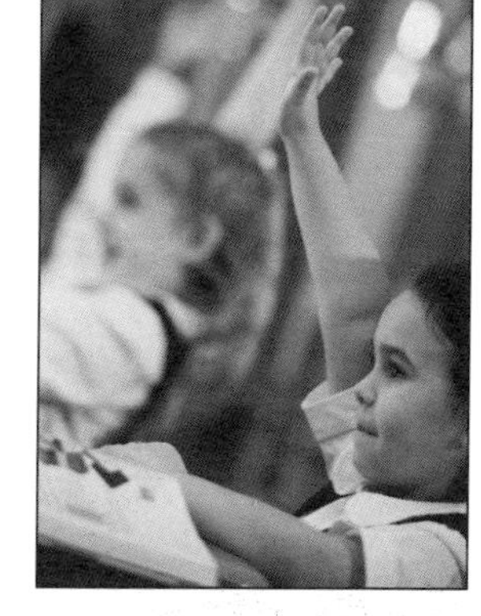

Alternatively, one can always turn the question around and ask what the child thinks. Many of the young child's questions are partly rhetorical in the sense that he or she has thought about them and may well have come up with his or her own answers. The child is most happy to share these thoughts with us. If we accept these answers without challenging them, we get insight into the child's thinking and communicate that we are interested in his or her ideas, not in right or wrong answers. In so doing we are not reinforcing wrong answers so much as promoting the child's sense of self-confidence and security in expressing his or her ideas.

There are limits to what one can effectively teach young children in the fields of math, science, and technology. But there are no limits to the young child's curiosity and imagination if we support and encourage his or her own ways of thinking.

### *Employing Motivation in Math, Science, and Technology Education*

Young children, as noted earlier in this paper, have their own curriculum priorities. They also have their own motivations for assigning these priorities. Individuals who have worked with young children remark upon how eager children are to explore and learn about their world. The origins of this motivation, however, are a matter of dispute. Some argue that children are intrinsically motivated to learn and that it is rigid, constraining schooling that dulls the young child's avid striving to learn about his or her world. Piaget (1950), in contrast, suggests that the spontaneous motivation we observe is actually a byproduct of developing mental abilities. When a child's abilities are maturing, he or she spontaneously seeks out stimuli to nourish those abilities. Young children will often ask adults to give them numbers to add or subtract as they begin to master these mathematical concepts.

If we accept this relationship between motivation and mastering concepts, it means that once the skills are fully acquired, the motivation will be lost. Indeed, this is what seems to happen during the early grades of school. It is not that the school deadens motivation, but rather that the spontaneous motivation associated with developing mental structures has dissipated. What takes the place of that spontaneous motivation is social motivation. Parents who are curious—who read newspapers, magazines, and books and who talk about the events of the world—encourage their children to be curious. Children, in turn, believe that by following the parental example they will please their parents and warrant their continued love and protection. In early childhood education, as in later education, parental modeling is all-important. Involving parents is an important part of effective early education.

Before leaving the subject of motivation, it is important to distinguish between capacity and learning, particularly since our knowledge about brain growth is increasing as are the number of facile extrapolations to early childhood education. As John T. Bruer (1997) has made very clear, such extrapolations are quite premature and reflect a failure to appreciate the complexities and intricacies of brain growth. Even if the young brain is growing rapidly, that does not tell us what type and how much stimulation is most conducive to productive learning. As Jane Healy wrote:

> Unproven technologies...may offer lively visions, but they can also be detrimental to the development of the young plastic brain. The cerebral cortex is a wondrously well-buffered mechanism that can withstand a good bit of well-intentioned bungling. Yet there is a point at which fundamental neural substrates for reasoning may be jeopardized for children who lack proper physical, intellectual and emotional nurturance. Childhood—and the brain—have their own imperatives. In development missed opportunities may be difficult to recapture. (Healy 1991)

In addition to inappropriateness, there is also the issue of the relationship of motivation to capacity. A case in point is the young child's great facility for learning foreign languages. It is generally accepted that the early years are the time to acquire a second language. Children who learn a second language early often speak without accent and with the rhythm and intonation of a native speaker. This unquestioned capacity of young children to learn a second language has prompted some parents and schools to provide second-language learning in kindergarten and first grade. Around the country, young children are being instructed in languages from French to Japanese. Much of this instruction is wasted time and effort.

Learning a foreign language requires more than capacity, it requires motivation. If a child has parents or grandparents who speak a foreign language, the child has plenty of motivation to learn that language. But there is no such motivation if the child is just given lessons in the language. He or she does not need it for anything and cannot use it for anything. Even with the capacity to learn a foreign language, without the motivation, that capacity will not be realized. It is not unlike the natural athlete who lacks the competitive zeal to succeed in a sport. Another individual, with less native ability but more ambition, may succeed where the natural athlete failed. It is when motivation and capacity work together that we find the most successful results.

## Conclusion

In this paper I have suggested that there are three major obstacles to effective math, science, and technology instruction in early childhood. The first barrier is the futility of using reflective or logical analysis as a means of arriving at how young children learn. The second obstacle is the transductive nature of young children's thinking. A final barrier resides in the fact that young children have their own curriculum priorities and their own motivations, which may be different than our own. These obstacles are not insurmountable but must be addressed to engage in meaningful and effective early childhood education in math, science, and technology.

A few strategies for overcoming these roadblocks were suggested. To overcome the barrier of learning how children learn, we have to observe children learning. To surmount the barrier of the limits of children's transductive thinking, we need to encourage their unlimited imagination and curiosity. Finally, we need to engage children's spontaneous motivation. But we also need to help instill social motivation in children by involving parents in ways that encourage their children to read, ask questions, and gather knowledge.

Early childhood is a most important period for math, science, and technology education, but only if we adapt such instruction to the unique needs, interests, and abilities of young children.

## References

Bruer, J.T. (1997, November). Education and the brain: A bridge too far. *Educational Researcher*, 4–16.

Bruner, J. (1962). *The process of education*. Cambridge, MA: Harvard University Press.

Elkind, D. (1987). *Miseducation*. New York: Knopf.

Froebel, F. (1904). *Pedagogics of the kindergarten* (J. Jarvis, Trans.). New York: D. Appleton & Company.

Healy, J. (1991). *Endangered minds. Why our children don't think*. New York: Touchstone Books.

Inhelder, B., and Piaget, J. (1958). *The growth of logical thinking from childhood to adolescence*. New York: Basic Books.

Kuhn, T. (1962/1970). *The structure of scientific revolutions*. Chicago: University of Chicago Press.

Piaget, J. (1942). *The child's conception of number*. London: Routledge & Kegan Paul.

Piaget, J. (1950). *The psychology of intelligence*. London: Routledge & Kegan Paul.

# First Experiences in Science, Mathematics, and Technology

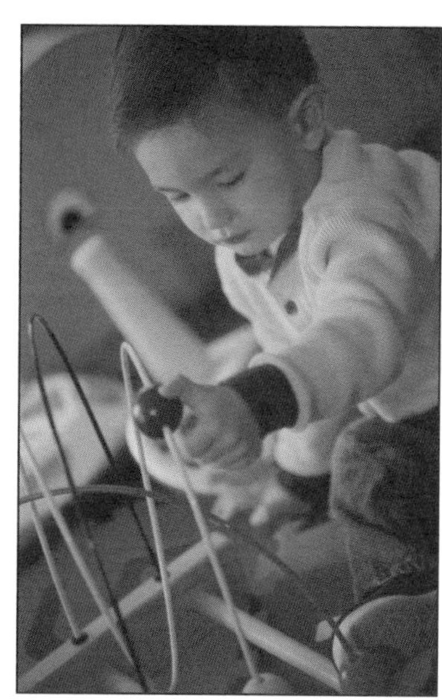

# Science in Early Childhood: Developing and Acquiring Fundamental Concepts and Skills

*Karen K. Lind*

One of the strongest themes in the *National Science Education Standards* (*NSES*) (National Research Council 1996) and *Benchmarks for Science Literacy* (*Benchmarks*) (American Association for the Advancement of Science 1993) is that all children can learn science and that all children should have the opportunity to become scientifically literate. In order for this learning to happen, the effort to introduce children to the essential experiences of science inquiry and explorations must begin at an early age.

A national consensus has evolved around what constitutes effective science teaching and learning for young children. More than ever before, educators agree that preschool-level and primary-level science is an active enterprise. Science is understood to be a process of finding out and a system for organizing and reporting discoveries. Rather than being viewed as the memorization of facts, science is seen as a way of thinking and trying to understand the world. This agreement can be seen in the national reform documents *NSES*, *Benchmarks*, and *Science for All Americans* (American Association for the Advancement of Science 1989.) Both *NSES* and *Benchmarks* are aligned with the guidelines from the National Association for the Education of Young Children (Bredekamp 1987; Bredekamp and Rosegrant 1992; Bredekamp and Copple 1997).

The reform documents mentioned in the previous paragraph espouse the idea that active, hands-on, conceptual learning provides meaningful and relevant learning experiences. These documents also reinforce Oakes' (1990) observation that all students, especially those in underrepresented groups, need to

*Karen K. Lind is an associate professor of science education at the University of Louisville.*

learn scientific skills such as observation and analysis at a very young age.

This paper describes how fundamental concepts and skills are developed from infancy through the primary years and offers strategies for helping students to acquire the skills needed for inquiry learning. It provides an overview of teaching and learning science in the early years, emphasizing the importance of selecting science content that matches the cognitive capacities of students.

## How Fundamental Concepts and Skills Develop

As any scientist knows, the best way to learn science is to do science. This is the only way to get to the real business of asking questions, conducting investigations, collecting data, and looking for answers. With young children, this strategy can best be accomplished by examining natural phenomena that can be studied over time. Children need to have a chance to ask and answer questions, do investigations, and learn to apply problem-solving skills. Active, hands-on, student-centered inquiry is at the core of good science education.

Concepts are the building blocks of knowledge; they allow people to organize and categorize information. During early childhood, children actively engage in acquiring fundamental concepts and in learning fundamental process skills. As we watch children in their everyday activities at various stages of development, we can observe them constructing and using concepts such as

- *one-to-one correspondence*—putting pegs in pegboard holes or passing one apple to each child at the table;
- *counting*—counting the pennies from the penny bank or the number of straws needed for every child at the table;
- *classifying*—placing square shapes in one pile and round shapes in another or putting cars in one garage and trucks in another; and
- *measuring*—pouring sand, water, rice, or other materials from one container to another.

Young children begin to construct many concepts during the pre-primary period, including mathematics and science concepts. They also develop the processes that enable them to apply their newly acquired concepts, expand existing concepts, and develop new ones. As they enter the primary period (grades one through three), children apply their early, basic concepts when exploring more abstract inquiries and concepts in science. Using these concepts also helps them understand more complex concepts in mathematics such as multiplication, division, and the use of standard units of measurement (Charlesworth and Lind 1995).

Concepts used in science grow and develop as early as infancy. Babies explore the world with their senses. They look, touch, smell, hear, and taste. Children are born curious and want to know all about their environment. As children learn to crawl, to stand, and to walk, they are free to discover more on their own and learn to think for themselves. They begin to learn ideas of *size*: As they look about, they sense their relative smallness. They go over, under, and into large objects and discover the size of these objects relative to their own size. They grasp things and find that some fit their tiny hands, and others do not. Infants learn about *weight* when they cannot always lift items of the same size. They learn about *shape*: Some things stay put while others roll away. They learn *time sequence*: When they wake up, they feel wet and hungry. They cry. The caretaker comes. They are changed and then fed. Next they play, get tired, and go to sleep. As babies first look and then move, they discover *space*: Some spaces are big and some spaces are small. With time, babies develop *spatial sense*: They are placed in a crib or playpen in the center of the living room (Charlesworth and Lind 1995).

Toddlers sort things. They put them in piles—of the same color, the same size, the same shape, or with the same use. Young children pour sand and water into containers of different sizes. They pile blocks into tall structures and see them fall and become small parts again. The free exploring and experimentation of a child's first two years help to develop muscle coordination and the senses of taste, smell, sight, and hearing—skills and senses that serve as a basis for future learning.

As children enter preschool and kindergarten, exploration continues to be the first step in dealing with new situations. At this time, however, children also begin to apply basic concepts to *collecting and organizing data* to answer a question. Collecting data requires skills in observation, counting, recording, and organizing. For example, for a science investigation, kindergartners might be interested in the process of plant growth. Supplied with lima bean seeds, wet paper towels, and glass jars, the children place the seeds in the jars, securing the seeds to the sides of the jars with the paper towels. Each day they add water, if needed, and observe what is happening to the seeds. They dictate their observation to their teacher, who records their comments on a chart. Each child also plants some beans in dirt in a small container such as a paper or plastic cup. The

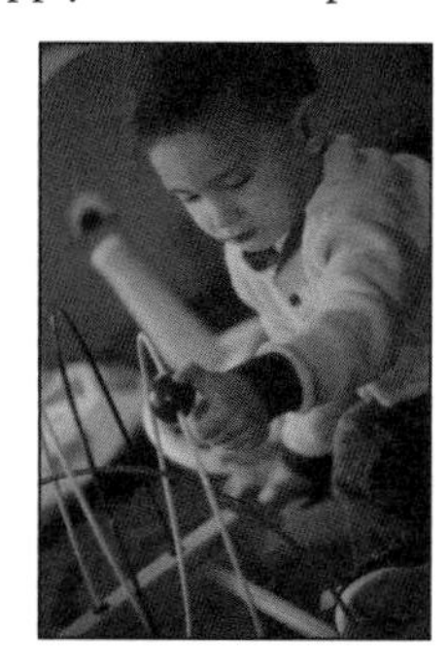

teacher supplies each child with a chart for his or her bean garden. The children check off each day on their charts until they see a sprout. Then they count how many days it took for a sprout to appear, comparing this number with those of other class members, as well as with the time it takes for the seeds in the glass jars to sprout. The children have used the concepts of number and counting, one-to-one correspondence, time, and comparison of the number of items in two groups. Primary-level children might attack the same problem, but they can operate more independently and record more information, use standard measuring tools, and do background reading on their own.

## How Science Concepts Are Acquired

Children acquire fundamental concepts through active involvement with their environment. As they explore their surroundings, they actively construct their own knowledge. Charlesworth and Lind (1995) characterize specific learning experiences with young children as *naturalistic* (or spontaneous), *informal*, or *structured*. These experiences differ in terms of who controls the activity: the adult or the child. *Naturalistic experiences* are those in which the child controls choice and action; in *informal experiences*, the child chooses the activity and action, but adults intervene at some point; and in *structured experiences*, the adult chooses the experience for the child and gives some direction to the child's action. Keep in mind that there are variations in learning styles among groups of children and among different cultural groups. Thus, science content should be introduced when it is appropriate to do so, as illustrated in the following examples.

### *Naturalistic Experiences*

Naturalistic experiences are those initiated spontaneously by children as they go about their daily activities. These experiences are the major mode of learning for children during the sensorimotor period. Naturalistic experiences can also be a valuable mode of learning for older children.

With naturalistic experiences, the adult's role is to provide an interesting and rich environment for the child. That is, adults should offer many things for the child to look at, touch, taste, smell, and hear. The adult should observe the child's activity, note how it is progressing, and then respond with a glance, a nod, a smile, or a word of praise to encourage the child. The child needs to know when he or she is doing the appropriate things. Below are some examples of naturalistic experiences.

- Tamara takes a spoon from the drawer and says, "This is big." Mom says, "Yes."

- Cindy (age 4) sits on the rug sorting colored rings into plastic cups.
- Sam (age 5) is painting. He puts down a dab of yellow. Then he dabs some blue on top. "Hey! I've got green now," he exclaims.

### *Informal Learning Experiences*

The adult initiates informal learning experiences as the child is engaged in naturalistic experiences. These experiences are not preplanned: They occur when the adult's experience or intuition or both indicate that it is time to act. For example, the child might be on the right track in solving a problem but needs a cue or encouragement. In another situation, the adult might take advantage of a teachable moment to reinforce certain concepts. Some examples of informal experiences follow.

- "I'm six years old," says three-year-old Kate while holding up three fingers. Dad says, "Let's count those fingers. One, two, three fingers. You are three years old."
- Juanita (age 4) has a bag of cookies. Mrs. Ramirez asks, "Do you have enough for everyone?" Juanita replies, "I don't know." Mrs. R. asks, "How can you find out?" Juanita says, "I don't know." Mrs. R. replies, "I'll help you. We'll count them."

### *Structured Learning Experiences*

Structured experiences are preplanned lessons or activities that can occur in many different ways. For example, Cindy is four years old. Her teacher decides that she needs to practice counting. She says, "Cindy, I have some blocks here for you to count. How many are in this pile?"

Teachers can also offer structured experiences in the following situations:

- With a small group at a specific time. For example, a teacher shows the children balls of different sizes and asks them to examine the balls and discuss their characteristics. The teacher picks up a ball and says, "Find a ball that is smaller."
- At any opportune time. Mrs. Flores, knowing that Tanya needs help with the concept of shape, suggests a game to play and gives her directions to play the game.
- With a large group at a specific time. Ms. Hebert realizes that classification is an important concept that should be applied throughout the primary grades. It is extremely important in organizing science data. For example, when it was time to study skeletons, Ms. Hebert had students bring bones from home so they could classify them.

## Commonalities of Science and Mathematics in Early Childhood

There is a natural integration of fundamental concepts and process skills across content areas, including mathematics and science. When fundamental mathematics concepts—comparing, classifying, and measuring—are applied to science problems, they are referred to as *process skills*. These mathematical concepts are necessary to solve some science problems. The other science process skills—observing, communicating, inferring, hypothesizing, and defining and controlling variables—are equally important for solving problems in both science and mathematics.

For example, consider the principle of the ramp, a basic concept in physics. Suppose a two-foot-wide plywood board is leaned against a large block, so that it becomes a ramp. Children are given a number of balls of different sizes and weights to roll down the ramp. Once their free exploration defines the ideas of the game, the teacher might ask some questions such as, "What would happen if two balls started to roll from the top of the ramp at the same time?" "What would happen if you changed the height of the ramp? Or had two ramps of different heights? Of different lengths?" The children could guess, explore what happens when they vary the steepness and length of the ramps or use different balls, observe what happens, communicate their observations, and describe similarities and differences in each of their experiments. They might observe differences in speed and distance contingent on the size or weight of the ball, the height and length of the ramp, or other variables. In this example, children could use the mathematical concepts of speed, distance, height, length, and counting (how many blocks are supporting each ramp?) while engaged in scientific observations.

In another example, a preschool teacher brings several pieces of fruit to class: one red apple, one green apple, two oranges, two grapefruit, and two bananas. The children examine the fruit to discover as much about these pieces as possible. They observe size, shape, color, texture, taste, and composition using counting and classification skills. (How many of each fruit type? Juicy or dry? Segmented or whole? Seeds or no seeds?) These observations may be recorded. (What is the color of each fruit? How many are spheres? How many are juicy?) The fruit can be weighed and measured, prepared for eating, and divided equally among the students.

Math and science concepts and skills can be acquired as children engage in traditional early childhood activities such as playing with blocks, water, sand, and manipulative materials, as well as during dramatic play, cooking, and outdoor activities. Providing young children with opportunities to see the math and science in their everyday activities helps them to build the basic understandings and interest for future learning.

## Encouraging Inquiry Through Problem Solving

A major area of interest in science education research is the teaching of science through inquiry. Research findings and the national reforms in science education overwhelmingly support this notion. The U.S. Department of Education and the National Science Foundation (1992) endorse mathematics and science curricula that promote active learning, inquiry, problem solving, cooperative learning, and other instructional methods that motivate students. The publication entitled *National Science Education Standards* (National Research Council 1996) states that science teaching must reflect science as it is practiced and that one goal of science education is to prepare children to understand and use the modes of reasoning of scientific inquiry. *NSES* presents inquiry as a step beyond process that involves learning, observing, and inferring.

Inquiry-oriented instruction engages students in the investigative nature of science. As Novak (1977) suggested, inquiry is a student behavior that involves activity and skills, but the focus is on the active search for knowledge or understanding to satisfy students' curiosity. In inquiry, educators should not expect children to discover everything for themselves, rather, they should focus on relating new science knowledge both to previously learned knowledge and to experiential phenomena, so students can build a consistent picture of the physical world. Science teachers can facilitate this process in several ways. For example, when children show an interest in learning more about a bean plant or a nearby tree, the teacher should ask questions to determine what each student already knows. In this way, teachers can modify learning experiences and classroom settings to best meet individual needs.

One way to involve students in inquiry is through problem solving, which is not as much a teaching strategy as it is a child behavior. As with inquiry, the driving force behind problem solving is curiosity—an interest in finding out. The challenge for the teacher is to create an environment in which problem solving can occur.

Problems should relate to, and include, the children's own experiences. From birth, children want to learn and they naturally seek out problems to solve. Problem solving in the pre-kindergarten years focuses on naturalistic and informal learning: filling and emptying containers of water, sand, or other substances; observing ants; or racing toy cars down a ramp. In kindergarten and the primary grades, adults can institute a more structured approach to problem solving.

Most science educators agree that problem solving and reflective thinking play an important role in children's science learning in school. In summarizing the findings of 26 national reports calling for reform in education—particularly curriculum and instruction in mathematics and science—Hurd (1989) found that 18 of those reports specifically identify problem solving in science as an educational objective.

Problem solving can be a powerful motivating factor to learn science. When students perceive the situations and problems they study in class as real, their curiosity is piqued and they are inspired to find an answer. Searching for a solution to a question or problem that is important to the student holds his or her attention and creates enthusiasm.

## The Theoretical Basis of Science Education

The young child's understanding of science grows from the fundamental concepts they develop during early childhood. Much of our understanding about how and when this development takes place comes from research that is based on theories of concept development as put forth by Jean Piaget and Lev Vygotsky (DeVries and Kohlberg 1987/1990; Driver et al. 1985; Kamii and DeVries 1978; Osborne and Freyberg 1985). These theories gave rise to the constructivist approach, which places the emphasis on individual children as intellectual explorers who make their own discoveries and construct knowledge. Constructivism has important implications for science education, especially in today's classrooms, where students are encouraged to engage in the inquiry process rather than memorize isolated science facts.

The current interest in the study of science concept learning owes much to the work of Novak (1977), whose book explores children's explanations for natural phenomena. Since this text was published, numerous studies related to a wide range of topics in the science curriculum have been reported, reviewed, and summarized by many researchers.

In science, teaching for conceptual change, or "teaching for understanding," requires different strategies from those previously employed by educators. Many science education researchers agree that the key is to provide a developmentally appropriate context that progressively increases in conceptual depth and complexity as children advance through school and life. The assessment of prior knowledge is thought to be essential to this process. Von Glasserfeld (1989), Resnick (1987), and others caution that if we as educators do not take students' prior knowledge into consideration, it is likely that the message we think we are sending will not be the message that students receive.

## Science Content and Cognitive Capacity: Avoiding a Mismatch

Although Piaget's (1969) developmental stages of learning are considered a major contribution to the teaching and learning of science, educators and curriculum developers do not always take these stages into account when designing science curriculum and experiences for young children. If children are to learn science and become scientifically literate, educators must choose appropriate science content and experiences to match children's cognitive capacities at different stages of their development.

Cowan (1978) underscores the importance of this alignment, stressing that mismatching content and developmental levels (e.g., expecting kindergarten children to understand the movements of the Earth's crust) leads to misconceptions and frustrations for teacher, parent, and child. These types of mismatches often cause teachers to resort to telling the information in a didactic manner because the child cannot conceptualize the content. As Covington and Berry (1976) found, the results of mismatched content and cognitive capacity are that (1) children are not able to extend, apply, or interpret deeper meanings of the content; and (2) interest and positive attitudes toward science are likely to diminish. Many other examples in the literature also emphasize the match between science content and cognitive capacity as essential to learning science. The implication from the research is that the content must always be within the realm of possibility of comprehension.

A prominent feature of cognitive research is the study of student misconceptions in science. These misconceptions are not merely errors in calculations or the misapplication of strategies. They are ideas that are based on misperceptions or incorrect generalizations that are consistent with the student's general understanding of a phenomenon. For example, misconceptions can be seen in children's ideas about light and shadows, which have been studied by Piaget (1930) and Feher and Rice (1987). Young children think of a shadow as an object.

They think that light is the agent that causes the object to form or that allows people to see the shadow, even when it is dark. This example clearly shows that misconceptions are a very real and significant obstacle to learning, one that educators must overcome before broaching new science concepts.

In considering all of the preschool and primary developmental stages described by Piaget, keep in mind that a child's view of the world and of scientific and mathematical concepts is not the same as yours. Their perception of phenomena is formed from their own perspective and experiences. Misconceptions will arise. So, be ready to explore the world to expand their thinking, and be prepared for the next developmental stage. Teach children to observe with all of their senses and to classify, predict, and communicate, so they can discover other viewpoints.

## References and Bibliography


American Association for the Advancement of Science. (1989). *Science for all Americans*. New York: Oxford University Press.

American Association for the Advancement of Science. (1993). *Benchmarks for science literacy*. New York: Oxford University Press.

Bredekamp, S., ed. (1987). *Developmentally appropriate practice in early childhood programs serving children from birth through age eight*. Washington, DC: National Association for the Education of Young Children.

Bredekamp, S., and Copple, C., eds. (1997). *Developmentally appropriate practice in early childhood programs: Revised.* Washington, DC: National Association for the Education of Young Children.

Bredekamp, S., and Rosegrant, T. (1992). *Reaching potentials: Appropriate curriculum and assessment for young children (Vol. 1).* Washington, DC: National Association for the Education of Young Children.

Charlesworth, R., and Lind, K. (1995). *Math and science for young children*. 2d ed. Albany, NY: Delmar.

Covington, M., and Berry, R. (1976). *Self-worth and school learning*. New York: Holt, Rinehart & Winston.

Cowan, P.A. (1978). *Piaget with feeling*. New York: Holt, Rinehart & Winston.

DeVries, R., and Kohlberg, L. (1987/1990). *Constructivist early education: Overview and comparison with other programs.* Washington, DC: National Association for the Education of Young Children.

Driver, R., Guesne, E., and Tiberghein, A., eds.. (1985). *Children's ideas in science*. Philadelphia, PA: Open University Press.

Feher, E., and Rice, K. (1987). Shadows and anti-images. *Science Education*, 725: 637–49.

Hurd, P.D. (1989). *Science education and the nation's economy*. Paper presented at the American Association for the Advancement of Science Symposium on Science Literacy, Washington, DC.

Kamii, C., and DeVries, R. (1978). *Physical knowledge in preschool education: Implications of Piaget's theory*. Englewood Cliffs, NJ: Prentice Hall.

Lind, K.K. (1997). Science in the developmentally appropriate integrated curriculum. In *Integrated curriculum and developmentally appropriate practice*, eds. D. Burts, C. Hart, and R. Charlesworth. Albany, NY: State University of New York.

Lowery, L.F. (1992). *The scientific thinking process*. Berkeley, CA: Lawrence Hall of Science.

National Research Council. (1996). *National science education standards*. Washington, DC: National Academy Press.

Novak, J. (1977). *A theory of education*. Ithaca, NY: Cornell University Press.

Oakes, J. (1990). *Lost talent: The under-participation of women, minorities, and disabled persons in science*. Santa Monica, CA: The Rand Corporation.

Osborne, M., and Freyberg, P. (1985). *Learning in science: Implications of children's science*. Auckland, New Zealand: Heinemann.

Piaget, J. (1930). *The child's conception of physical causality*. Totowa, NJ: Littlefield, Adams.

Piaget J. (1969). *Psychology of intelligence*. Totowa, NJ: Littlefield, Adams.

Resnick, L.B. (1987). *Education and learning to think*. Washington, DC: National Academy Press.

United States Department of Education and National Science Foundation. (1992). *Statement of principles*. (Brochure). Washington, DC: Author.

Von Glasserfeld, E. (1989). Cognition, construction of knowledge, and teaching. *Syntheses*, 80: 121–40.

# Early Childhood Mathematics

*Susan Sperry Smith*

A group of three-year-olds sits in a circle, eagerly awaiting the story, *Puppy Says 1-2-3* (Singer 1993). They have not heard the tale of the puppy that counts and squeaks as their teacher, Miss Lily, squeezes his tummy. The class counts along with puppy. "Puppy looks up and what does he see? ONE little squirrel climbing up ONE big tree. Puppy says ONE."

Then the children play a game with three rubber cows, a mat of green "grass," and a bowl turned into a barn. One child says, "Let's put one cow on the grass. Let's put another cow on the grass. Now we have *two* cows." The cows go *into* the barn, *around* the barn, and *on top of* the barn. The teacher gives them makeshift cardboard "bridges." The cows go *over* the bridge and *under* the bridge.

*Susan Sperry Smith is an associate professor in the College of Education at Cardinal Stritch University in Milwaukee, Wisconsin.*

The teacher prepares many centers that are found throughout the room. Besides housekeeping, a picture-book center, a computer center, and a puzzle center, there are many centers devoted to early mathematical experiences. After much repetition, the puppy book will go in the picture-book center to be read and reread. While the teacher knows that some of the children have been able to count to 10 since the age of two, she realizes that very young children need a foundation with small sets, 1-2-3, emphasizing one-to-one correspondence. At a later time Miss Lily will introduce numbers up to five or 10.

During their music time, the class sings very simple songs with much repetition. Today they sing "Row, Row, Row Your Boat." The sound of the music and the lyrics reinforce the concept of pattern. Later in the year the children will construct simple patterns in artwork and with manipulatives.

## Mathematical Activities

The children choose a center to explore. The matching center contains bins of socks, mittens, identical farm animals, and zoo animals. The children put identical pairs together by type, not color. In the comparing center, a child sorts stuffed animals into "big" piles and "little" piles with the teacher's help. Next week, the comparing center will feature a different pair of words.

Later in the week, the class gathers in a circle to practice sorting. They concentrate on things they know, such as food, toys, clothes, and ways to travel. They sort themselves by boy/girl, hair clips/no hair clips, buttons/no buttons, and so on. Miss Lily avoids categories that might cause hard feelings, such as a certain brand of tennis shoe versus a dress shoe.

In another session, Miss Lily introduces nesting toys that illustrate ordering. She asks the children to find the biggest one. The children point to the biggest cup. She takes it out of the line, places it near her, and asks, "Now which one is the biggest?" Miss Lily gathers various nested sets of measuring cups, kitchen bowls, plastic glasses, and commercial nesting toys for her ordering center. Throughout the week, children choose to visit this center, and they try to put the items in order of size.

In the pouring center, a child fills plastic containers of many sizes with scoops of rice. The teacher helps with words like *empty/full* and *little/big*. The child fills the cup to the *top*. Later in the year, the class will discuss "which jar has *more*, and which jar has *less*." A sturdy balance sits on the counter nearby. Children take turns weighing fruit, feathers, and small items such as erasers, chalk, markers, crayons, and toy cars. Each child tells Miss Lily which items are *heavy* and which are *light*.

The block-building center is a major center and the cornerstone of a mathematically rich environment. Blocks are essential tools for creativity, dramatic play, and geometry—for girls as well as boys. The teacher rotates groups of children to give everyone a chance to use the building blocks. At first, a child might pile the blocks to make a tower, then make a simple enclosure, and eventually master the challenge of roofing or bridging the space between two walls. With time and practice, children may build elaborate structures that have evidence of symmetry, sound construction, and aesthetics.

### *Developing Spatial Sense*

Miss Lily understands that developing concepts about space is a natural part of growing up. She recognizes that children need opportunities to study the relationships between objects, places, and events (the study of topology) more than they need the ability to draw common shapes such as a circle or a square.

Miss Lily creates opportunities to explore *proximity*, asking questions such as "Where am I?" or "Where is it?" *Separation* refers to the ability to see the whole object as comprised of individual parts. Puzzles and model building encourage this ability. The nesting-center toys promote *order*, including reversing one's thinking. Miss Lily also talks about last week's events as well as what is happening today.

*Enclosure* refers to being surrounded or boxed in by the surrounding objects. The points on either side can enclose a point on a line. In three-dimensional space, a fence can enclose the animals, or a canister with a lid can enclose the cereal. The teacher helps by saying, "Is the lid closed so the beads won't spill out?" or "Open the closed door so we can hang up our coats."

All of these activities contribute to the overall development of a child's *spatial* sense. In its 1989 *Curriculum and Evaluation Standards for School Mathematics*, the National Council of Teachers of Mathematics (NCTM) defines spatial sense this way:

> Spatial sense is an intuitive feel for one's surroundings and the objects in them. To develop spatial sense, children must have many experiences that focus on geometric relationships: the direction, orientation, and perspectives of objects in space, the relative shape and sizes of figures and objects, and how a change in shape relates to a change in size. (page 49)

Spatial sense contributes to the study of geometry, and is an integral part of the preschool curriculum.

### *Developing Number Sense*

*Number sense* is using common sense based on the way numbers and tools work within a given culture. It involves an appreciation for the reasonableness of an answer and the level of accuracy needed to solve a particular problem. It is a complex set of interrelated concepts (Smith 1997), including

- reading numerals (For example, "It's a three.")
- writing numerals, a visual-motor task.
- matching a number to a set, or the principle of

cardinality. (A child counts five beans and answers the question, "How many?")

- having an intuitive feel for how big a number is. ("Is 15 closer to 10 or closer to 50?")
- being able to make reasonable guesses using numbers. (The small jar could not hold more than 100 goldfish crackers.)
- seeing part-whole relationships using sight or abstract thinking, not counting. ("I have two green bottle caps and three purple bottle caps.")

The teacher facilitates the development of number sense and spatial sense throughout the preschool years.

## The Teacher's Role

Most experts believe that young children possess a substantial amount of informal knowledge about mathematics. The teacher's role is to create a link between children's ability to use informal math and the ability to understand the more formal math found in grade school (Ginsberg 1996).

Teachers must help children construct and elaborate upon what they already know, so they can "re-invent" mathematics for themselves. A reflective teacher helps the child discover and communicate ideas that would have not occurred spontaneously to the child without the adult's help (Vygotsky 1978). As children mature, they find patterns and solve problems far beyond what is typically found in the preschool-kindergarten curriculum (Resnick et al. 1991; Carpenter et. al. 1993).

## The Kindergarten Program

Mr. Toby has a kindergarten class of 15 students. At the beginning of each day, they place a picture of themselves that has been glued to a magnetized orange juice lid on the attendance chart. The children count and decide how many people are in class today and how many are absent. They decide if there are more boys than girls or vice versa, and they figure out the difference. They chart the weather for the day by placing a magnetized counter under the category chosen: sun, clouds, rain, snow. They know that they can only choose one at a certain time in the morning.

They sequence the day's activities in a pocket chart. Mr. Toby knows that young children cannot comprehend the traditional calendar, i.e., a five-row and seven-column matrix, with both ordinal and cardinal numbers (Schwartz 1994). He will gradually use a weekly schedule and then a two-week schedule before introducing a more comprehensive calendar. The children keep track of how many days they have been in school by putting a straw for each day in a container labeled the "1's cup." When there are 10 straws in the cup, they bundle them and move them to the "10's cup." Sometime in February there will be 10 groups of 10 straws to move to the "100's cup," and the class will celebrate the 100$^{th}$ day of school. They will decorate a cake with 100 candles and have a party with a "GORP" mix, which consists of small snack items (raisins, cereal, chocolate chips, etc.) that the students bring from home and sort into groups of 100. They will enjoy a day filled with many activities that use 100 items.

The circus is a popular kindergarten theme. Over several days the class will participate in many creative art, creative movement, science, dramatic play, and cooking activities. A teacher plans a number of math activities, some of which are described in the text that follows.

### Math Activity 1: Number and Measurement

*Name of Activity*: Peanut Perimeter

*Materials*: Peanuts in the shell, a large bowl for each pair of students, small tables.

*Tasks*:

1. In pairs, the children decide how to line the edges of a small table with peanuts. They pay close attention to covering the edge and having the peanuts touch.
2. After they finish lining the perimeter of the table, the children remove the peanuts while counting them with the teacher. Kindergarten children enjoy counting to 100 and beyond. (It may take more than 100 peanuts to line the perimeter of some tables.)

### Math Activity 2: Sequence and Ordering (Time)

*Name of Activity*: Mirette's Story

*Materials*: The book, *Mirette on the High Wire* (McCully 1992).

*Tasks*:

1. Read and reread the story of Mirette, and highlight the events in sequence:
   a. Mirette lives in a boardinghouse.
   b. A new tenant, a retired high-wire performer, arrives.
   c. He teaches Mirette to walk the high wire.
   d. He returns to the stage.
2. Have the class act out and retell this story in sequence.

### Math Activity 3: Measurement and Weight

*Name of Activity*: How Much Does a Baby Elephant Weigh?

*Materials*: Pictures of things that are very heavy, such as a baby elephant, and pictures of things that are very light, such as poster board or bulletin board.

*Tasks*:

1. Research the weight of a baby elephant. Compare it to the weight of a newborn person.
2. Make a more weight/less weight chart, with pictures of things that might weigh more or less than a baby elephant.

### Math Activity 4: Part-Part-Whole—The Number 6

*Name of Activity*: Mixed Nut Designs

*Materials*: Nuts in the shell, such as peanuts, almonds, walnuts, and pecans (any nuts that do not roll); a bowl; a large table or a rug.

*Tasks*:

1. Make designs with two kinds of nuts, so each design uses six nuts.
2. Fill the whole table with designs. Tell the teacher about your combinations, for example, "This one has two pecans and four peanuts. It looks like a star."

(For additional math activities, see Smith 1997.)

Mr. Toby concentrates on pattern-work and part-part-whole designs with each number from 4 to 12. Later in the year, he will introduce simple story problems, following the Cognitively Guided Instruction Approach (Carpenter and Moser 1983; Carpenter and Moser 1984; Carpenter et al. 1990; Peterson et al. 1989). Many kindergartners are able to solve the following types of problems by using counters or their fingers or by drawing.

- *The circus ring had three clowns. Four more clowns join them. Now how many clowns are in the ring?*
- *The clown had nine pieces of candy. He gave away four pieces. How many pieces does the clown have left?*

Some kindergarten children can also solve simple multiplication (repeated addition) and simple division (repeated subtraction) problems. For example,

- *The clown had three bags of candy. There were five pieces of candy in each bag. How many pieces did the clown have?*
- *The clown had 12 pieces of candy. He gave three pieces to each child. How many children received candy?*

### The Teacher's Role

Mr. Toby's classroom provides the time and structure needed by children to explore significant mathematics. He encourages his students by respecting and valuing their ideas and validating their ways of thinking. He challenges the class to take intellectual risks by posing interesting questions to the group. They learn to support their responses with mathematical ideas.

Finally, he encourages all students to participate, so they gain confidence in their ideas.

All preschool and kindergarten teachers must pay attention to the key ingredients for success: a well-prepared environment, a developmentally appropriate math curriculum, and an awareness of the teacher's role. The process of learning is never over, but the journey is worth taking.

## References

Carpenter, T.P., and Moser, J.M. (1983). The acquisition of addition and subtraction concepts. In *The acquisition of mathematical concepts and processes*, eds. R. Lesh and M. Landeau, 7–44. New York: Academic Press.

Carpenter, T.P, and Moser, J.M. (1984). The acquisition of addition and subtraction concepts in grades one through three. *Journal for Research in Mathematics Education*, 15: 179–202.

Carpenter, T.P., Ansell, E., Franke, M.C., Fennema, E., and Weisbeck, L. (1993). Models of problem solving: A study of kindergarten children's problem-solving processes. *Journal for Research in Mathematics Education*, 24(5): 427–440.

Carpenter, T.P., Carey, D., and Kouba, U. (1990). A problem solving approach to the operations. In *Mathematics for the young child*, ed. J.N. Payne, 111–131. Reston, VA: National Council of Teachers of Mathematics.

Ginsberg, H.P. (1996). Toby's math. In *The nature of mathematical thinking*, eds. R.J. Sternberg and T. Ben-Zeev, 175–202. Hillsdale, NJ: Lawrence Erlbaum Associates.

McCully, E.A. (1992). *Mirette on the high wire*. New York: G.P. Putnam & Sons.

National Council of Teachers of Mathematics. (1989). *Curriculum and evaluation standards for school mathematics*. Reston, VA: Author.

Peterson, P., Fennema, E., and Carpenter, T. (1989). Using knowledge of how students think about mathematics. *Educational Leadership*, 46(4): 42–46.

Resnick, L., Bill, V., Lesgold, S., and Leer, N. (1991). Thinking in arithmetic class. In *Teaching advanced skills to at-risk students*, eds. B. Means, C. Chelmer, and M. Knapp. San Francisco: Jossey-Bass.

Schwartz, L.L. (1994). Calendar reading: A tradition that begs remodeling. *Teaching Children Mathematics*, 1: 104–109.

Singer, M. (1993). *Puppy says 1,2,3*. Hong Kong: Reader's Digest Young Families, Inc.

Smith, S.S. (1997). *Early childhood mathematics.* Needham Heights, MA: Allyn & Bacon.

Vygotsky, L.S. (1978). *Mind in society: The development of higher psychological processes.* Cambridge, MA: Harvard University Press.

# Young Children and Technology

*Douglas Clements*

Computers are increasingly present in early childhood education settings. Toward the end of the 1980s, only one-fourth of licensed preschools had computers. Today almost every preschool has a computer, with the ratio of computers to students changing from 1:125 in 1984 to 1:22 in 1990 to 1:10 in 1997. This last ratio matches the minimum ratio that is favorable to social interaction (Clements and Nastasi 1993; Coley et al. 1997). During the last 13 years, perspectives on the principle of developmental appropriateness have become more sophisticated. Researchers have extended these perspectives to include such dimensions as cultural paradigms and multiple intelligences (Bowman and Beyer 1994; Spodek and Brown 1993).

Research on young children and technology similarly has moved beyond simple questions to consider the implications of these changing perspectives for the use of technology in early childhood education. For example, we no longer need to ask whether the use of technology is "developmentally appropriate." Very young children have shown comfort and confidence in using software. They can follow pictorial directions and use situational and visual cues to understand and think about their activity (Clements and Nastasi 1993). Typing on the keyboard does not seem to cause them any trouble; if anything, it is a source of pride.

With the increasing availability of hardware and software adaptations, children with physical and emotional disabilities can also use the computer with ease. Besides enhancing their mobility and sense of control, computers can help improve their self-esteem. One totally mute four-year-old with diagnoses of retardation and autism began to echo words for the first time while working at a computer (Schery and O'Connor 1992). However, such access is not always equitable across our society. For example, children

*Douglas Clements is a professor in the department of learning and instruction in the graduate school of education at the SUNY–Buffalo.*

attending low-income and high-minority schools have less access to most types of technology (Coley et al. 1997).

Research has also moved beyond the simple question of whether computers can help young children learn. They can. What we need to understand is how best to aid learning, what types of learning we should facilitate, and how to serve the needs of diverse populations. In some innovative projects, computers are more than tools for bringing efficiency to traditional approaches. Instead, they open new and unforeseen avenues for learning. They allow children to interact with vast amounts of information from within their classrooms and homes. They tie children from across the world together (Riel 1994).

Not every use of technology, however, is appropriate or beneficial. The design of the curriculum and social setting are critical. This paper reviews the research in three broad areas: social interaction, teaching with computers, and curriculum and computers. Finally, it describes a new project that illustrates innovative, technology-based curriculum for early childhood education.

## Social Interaction

An early concern, that computers will isolate children, was alleviated by research. In contrast, computers serve as *catalysts* for social interaction. The findings are wide-ranging and impressive. Children at the computer spent nine times as much time talking to peers while on the computer than while doing puzzles (Muller and Perlmutter 1985). Researchers observe that 95 percent of children's talking during Logo work is related to their work (Genishi et al. 1985). (Logo is a computer programming language designed to promote learning. Even young children can use it to direct the movements of an on-screen "turtle.") Children prefer to work with a friend rather than alone. They foster new friendships in the presence of the computer. There is greater and more spontaneous peer teaching and helping when children are using computers (Clements and Nastasi 1992).

The software they use affects children's social interactions. For example, open-ended programs such as Logo foster collaboration. Drill-and-practice software, on the other hand, can encourage turn-taking but also competition. Similarly, video games with aggressive content can engender competitiveness and aggression in children. Used differently, however, computers can have the opposite effect (Clements and Nastasi 1992). In one study, a computer simulation of the playhouse from the animated t.v.

series *The Smurfs* attenuated the themes of territoriality and aggression that emerged with a real playhouse version of the Smurf environment (Forman 1986).

The physical environment also affects children's interactions (Davidson and Wright 1994). Placing two seats in front of the computer and one at the side for the teacher can encourage positive social interaction. Placing computers close to each other can facilitate the sharing of ideas among children. Centrally located computers invite other children to pause and participate in the computer activity. Such an arrangement also helps to keep teacher participation at an optimum level. Teachers are nearby to provide supervision and assistance as needed, but they are not constantly so close as to inhibit the children (Clements 1991).

## Teaching with Computers

The computer offers unique advantages in teaching. Opportunities to aid learning are addressed in the following section. Technology also offers unique ways to assess children. Observing the child at the computer provides teachers with a "window into a child's thinking process" (Weir et al. 1982). Research has also warned us not to curtail observations after a few months. Sometimes, beneficial effects appear only after a year. Ongoing observations also help us chart children's learning progress (Cochran-Smith et al 1988).

Differences in learning styles are more readily visible at the computer, where children have the freedom to follow diverse paths towards a goal (Wright 1994). This flexibility is particularly valuable with special children, as the computer seems to reveal their hidden strengths. Different advantages emerge for other groups of children. For example, researchers have found differences in Logo programming between African-American and Caucasian children. The visual nature of Logo purportedly was suited to the thinking style of African-American children's thinking style (Emihovich and Miller 1988).

Gender differences also emerge when children engage in programming. In one study, a post-test-only assessment seemed to indicate that boys performed better. However, assessment of the children's interactions revealed that the boys took greater risks and thereby reached the goal. In comparison, girls were more keen on accuracy; they meticulously planned and reflected on every step (Yelland 1994). Again, the implication for teaching is the need for consistent, long-term observation.

Yet another opportunity offered us by technology is to become pioneers ourselves. Because we know our children best, we can best create the program that will help them. Frustrated by the lack of good computer software, Tom Snyder started using the computer to support his classroom simulations of history. Mike Gralish, a first-grade teacher, used several computer devices and programs to link the base-10 blocks and the number system for his children. Today, both of these gentlemen are leading educational innovators (Riel 1994).

To be innovators and to keep up with the growing changes in technology, teachers need in-service training. Research has established that less than 10 hours of training can have a negative impact (Ryan 1993). Others have emphasized the importance of hands-on experience and warned against brief exposure to a variety of software programs, encouraging an in-depth knowledge of one program (Wright 1994).

## Curriculum and Computers

The computer also offers unique opportunities for learning through exploration, creative problem solving, and self-guided instruction. Realizing this potential demands a simultaneous focus on curriculum and technology innovations (Hohmann 1994). Effectively integrating technology into the curriculum demands effort, time, commitment, and, sometimes, even a change in one's beliefs.

We begin with several overarching issues. What type of computer software should be used? Drill-and-practice software leads to gains in certain rote skills. However, it has not been as effective in improving the conceptual skills of children (Clements and Nastasi 1993). Discovery-based software that encourages and allows ample room for free exploration is more valuable in this regard. However, research has shown that children work best with this type of software when they are assigned to open-ended projects rather than asked merely to "free explore" (Lemerise 1993). They spend more time and actively search for diverse ways to solve the task. The group of children who were allowed to free explore grew disinterested quite soon.

Another concern was that computers would replace other early childhood activities. Research shows that computer activities yield the best results when coupled with suitable off-computer activities. For example, children who are exposed to developmental software alone—the on-computer group—show gains in intelligence, non-verbal skills, long-term memory, and manual dexterity. Those who also worked with supplemental activities, in comparison—the off-computer group—gained in all of these areas and improved their scores in verbal, problem-solving, and conceptual skills (Haugland 1992). In addition, these children spent the least amount of time using the computers. A control group that used drill-and-practice software spent three times as long on the computer but showed less than half of the gains that the on- and off-computer groups did. Given these capabilities of the computer, how has it affected children's learning?

In mathematics specifically, the computer can provide practice on arithmetic processes and foster deeper conceptual thinking. Drill-and-practice software can help young children develop competence in counting and sorting (Clements and Nastasi 1993). However, it is questionable if the exclusive use of such drill-and-practice software would subscribe to the vision of the National Council of Teachers of Mathematics (NCTM) (1989): Children should be "mathematically literate" in a world where the use of mathematics is becoming more and more pervasive. NCTM recommends that we "create a coherent vision of what it means to be mathematically literate both in a world that relies on calculators and computers to carry out mathematical procedures and in a world where mathematics is rapidly growing and is extensively being applied in diverse fields" (National Council of Teachers of Mathematics 1989). This vision *de*-emphasizes rote practice on isolated facts. It emphasizes discussing and solving problems in geometry, number sense, and patterns with the help of manipulatives and computers.

For example, using software programs that allow the creation of pictures with geometric shapes, children have demonstrated growing knowledge and competence in working with concepts such as symmetry, patterns, and spatial order. Tammy overlaid two overlapping triangles on one square and colored select parts of this figure to create a third triangle that did not exist in the program! Not only did this preschooler exhibit an awareness of how she had made this figure, but she also showed awareness of the challenge it would be to others (Wright 1994). Using a graphics program with three primary colors, young children combined these colors to create three secondary colors (Wright 1994). Such complex combinatorial abilities are often thought to be out of the reach of young children. The computer experience led the children to explorations that expanded their boundaries.

Young children can also explore simple "turtle geometry." They direct the movements of a robot or screen "turtle" to draw different shapes. One group of five-year-olds was constructing rectangles. "I wonder if I can tilt one," mused one boy. He turned the turtle with a simple mathematical command, "L 1" (turn left one unit), drew the first side, then was unsure about what to do next. He finally figured out that he must use the same turn command as before. He hesitated again. "How far now? Oh, it *must* be the same as its partner!" He easily completed his rectangle. The instructions he should give the turtle *at this new heading* were, at first, not obvious. He analyzed the situation and reflected on the properties of a rectangle. Perhaps most important, he posed the problem for himself (Clements and Battista 1992).

This boy had walked rectangular paths, drawn rectangles with pencils, and built them on geo-boards and pegboards. What did the computer experience *add*? It helped him *link* his previous experiences to more explicit mathematical ideas. It helped him *connect* visual shapes with abstract numbers. It encouraged him to *wonder* about mathematics and pose problems in an environment in which he could create, experiment, and receive feedback about his own ideas.

Such discoveries happen frequently. One preschooler made the discovery that reversing the turtle's orientation and moving it backwards had the same effect as merely moving it forwards. The significance the child attached to this discovery and his overt awareness of it was striking. Although the child had done this previously with toy cars, Logo helped him abstract a new and exciting idea (Tan 1985).

## *Building Blocks©*: An Innovative Technology-Based Curriculum

At present, Julie Sarama and I are developing innovative pre-K to grade 2 curriculum materials. The project, "Building Blocks—Foundations for Mathematical Thinking, Pre-Kindergarten to Grade 2: Research-based Materials Development," is funded by a grant from the National Science Foundation. It is designed to enable all young children to build solid content knowledge and develop higher-order thinking. The program's design is based on current theory and research and represents a state-of-the-art technology curriculum for young children in the area of mathematics. It is discussed in this paper in that light. The reader might notice that our description does not begin with a listing of its technologically sophisticated elements, including multimedia features. This strategy is deliberate. We emphasize the art and science of teaching and learning, in contrast to many early childhood software programs, which use technologically advanced bells and whistles to disguise ordinary activities.

The design of a state-of-the-art curriculum must begin with audience considerations. The demographics of this age range imply that materials should be designed for home, daycare, and classroom environments and for children who have various backgrounds, interests, and ability levels. To reach this broad audience, the curriculum materials will be progressively layered: Users will be able to "dig deeper" into them to reach increasingly rich, but demanding, pedagogical and mathematical levels. The materials should not rely on technology alone. They should integrate three types of media: computers, manipulatives (and everyday objects), and print.

The project's basic educational approach is finding the mathematics in children's activities and developing mathematics from them. We focus on helping children extend and find mathematics in their everyday activities, from building blocks to art to songs to puzzles. Thus, we will design activities based on children's experiences and interests, with an emphasis on supporting the development of mathematical activity. This process emphasizes representation: using mathematical objects and actions that relate to children's everyday activities. Our materials will embody these actions-on-objects in a way that mirrors the theory of and research on children's *cognitive building blocks*—creating, copying, uniting, and dis-embedding both units and composite units.

Perhaps the most important aspect of the project's material design is our model for the design process. Curriculum and software design can and should have an explicit theoretical and empirical foundation, beyond its genesis in someone's intuitive grasp of children's learning. It should also interact with the ongoing development of theory and research, reaching toward the ideal of testing a theory by testing the software and the curriculum in which it is embedded. In this model, one conducts research at multiple aggregate levels, making the research relevant to educators in many positions. We have cognitive models with sufficient explanatory power to permit the design to grow co-jointly with the refinement of these cognitive models (Biddlecomb 1994; Clements and Sarama 1995; Fuson 1992; Hennessy 1995).

The phases of our nine-step design process model follow.

1. Draft curriculum goals.
2. Build an explicit model of children's knowledge and learning in the goal domain.
3. Create an initial design.
4. Investigate components.
5. Assess prototypes and curriculum.
6. Conduct pilot tests.
7. Conduct field tests in multiple settings.
8. Recurse.
9. Publish and disseminate.

These phases include a close interaction between materials development and a variety of research methodologies, from clinical interviews to teaching experiments to ethnographic participant observation.

A new technology permits the reflective consideration of objects, actions, and activities, which can help developers reconceptualize the nature and content of mathematics that might be learned. The developer can also conceive new designs by reflecting on how software might provide tools that enhance students' actions and imagination or that suggest an encapsulation of a process or obstacles that force students to grapple with an important idea or issue. Finally, the flexibility of computer technologies allows the creation of a vision less hampered by the limitations of traditional materials and pedagogical approaches (cf. Confrey, in press). For example, computer-based communication can extend the model for mathematical learning beyond the classroom. Computers can allow representations and actions not possible with other media. The materials in *Building Blocks* will not only ensure that computerized actions-on-objects mirror the goal concepts and procedures, but also that they are embedded in tasks and developmentally appropriate settings (e.g., narratives, fantasy worlds, building projects).

The materials will emphasize the development of basic *mathematical building blocks*—ways of knowing the world mathematically. These building blocks will be organized into two areas: (1) spatial and geometric competencies and concepts; and (2) numeric and quantitative concepts, based on the considerable research in that domain. Three mathematical sub-themes will be woven through both main areas: (1) patterns and functions; (2) data; and (3) discrete mathematics (e.g., classifying, sorting, sequencing). Most important will be the synthesis of these domains, each to the benefit of the other. The building blocks of the structure are not elementary school topics "pushed down" to younger ages; they are developmentally appropriate domains, that is, topics that are meaningful and interesting to children. Access to topics such as large numbers or geometric ideas such as depth, however, are not restricted. In fact, research indicates that these concepts are both interesting and accessible to young children.

By presenting concrete ideas in a symbolic medium, for example, the computer can help bridge these two concepts for young children. But are these manipulatives still "concrete" on the computer screen? One has to examine what *concrete* means. Sensory characteristics do not adequately define it (Clements and McMillen 1996; Wilensky 1991). First, it cannot be assumed that children's conceptions of the manipulatives are similar to those of adults (Clements and McMillen 1996). Second, physical actions with certain manipulatives may suggest different mental actions than those we wish students to learn. For example, researchers found a mismatch among students using the number line to perform addition. When adding five and four, the students located the number 5, counted "one, two, three, four" and read the answer. This action did not help them solve the problem mentally, for to do so they have to count "six, seven, eight, nine" and at the same time count the counts—6 is 1, 7 is 2, and so on. These actions are quite different (Gravemeijer 1991).

Thus, manipulatives do not always carry the meaning of the mathematical idea. Students must use these manipulatives in the context of well-planned activities and ultimately reflect on their actions in order to grasp the idea. Later, we expect them to have a "concrete" understanding that goes beyond these physical manipulatives.

It appears that there are different ways to define *concrete* (Clements and McMillen 1996). We define sensory-concrete knowledge as that in which students must use sensory material to make sense of an idea. For example, at early stages, children cannot count, add, or subtract meaningfully unless they have actual objects to aid in those functions. They build integrated-concrete knowledge as they learn. Such knowledge is connected in special ways. (The root of the word *concrete* is "to grow together.") What gives sidewalk concrete its strength is the combination of separate particles in an interconnected mass. What gives integrated-concrete thinking its strength is the combination of many separate ideas in an interconnected structure of knowledge (Clements and McMillen 1996).

For example, computer programs may allow children to manipulate on-screen "building blocks." These blocks are not physically concrete. However, no base-10 blocks "contain" place-value ideas (Kamii 1986). Students must build these ideas from working with the blocks and thinking about their actions. Furthermore, research indicates that physical base-10 blocks can be so clumsy and the manipulations so disconnected from each other that students see only the trees (manipulations of many pieces) and miss the forest (place-value ideas). Computer blocks can be more manageable and "clean" (Thompson and Thompson 1990). Students can break computer base-10 blocks into single blocks, or glue these blocks together to form 10s. These actions are more in line with the mental actions that we want students to learn: They are children's cognitive building blocks.

One essential cognitive "building block" of place value is children's ability to count by 10 from any number, thus constructing composite units of 10 (Steffe and Meinster 1997). The computer helps students make sense of their activity and the numbers by linking the blocks to symbols. For example, the number represented by the base-10 blocks is usually linked dynamically to the students' actions with the blocks, automatically changing the number spoken and displayed by the computer when the student changes the blocks. As a simple example, a child who has 16 single blocks might glue 10 together and then repeatedly duplicate this "10." In counting along with the computer, "26, 36, 46," and so on, the child constructs composite units of 10.

Computers encourage students to make their knowledge explicit, which helps them build integrated-concrete knowledge. Specific theoretically and empirically grounded advantages of using computer manipulatives follow (Clements and McMillen 1996).

- They provide a manageable, clean manipulative.
- They offer flexibility.
- They can change arrangement or representation.
- They can store, and later retrieve, configurations.
- They record and replay students' actions.
- They link the concrete and the symbolic with feedback.
- They dynamically link multiple representations.
- They change the very nature of the manipulative.
- They link the specific to the general.
- They encourage problem posing and conjecturing.
- They provide a framework for problem solving, focus attention, and increase motivation.
- They encourage and facilitate complete and precise explanations.

Of course, multimedia and other computer capabilities should, and will, be used when they serve educational purposes. Features such as animation, music, surprise elements, and especially consistent interaction get and hold children's interest (Escobedo and Evans 1997). They can also aid learning, if they are designed to support and be consistent with the pedagogical goals. In addition, access to technology is an important equity issue. Much of our material will be available on the Internet.

In summary, the *Building Blocks* project is designed to combine the art and science of teaching and learning with the science of technology, with the latter serving the former. Such synthesis of curriculum and technology development as a scientific enterprise with mathematics education research will reduce the separation of research and practice in mathematics and technology education. Materials based on research can then be produced, and research can be based on effective and ecologically sound learning situations. Moreover, these results will be immediately applicable by practitioners (parents, teachers, and teacher educators); administrators and policy makers; and curriculum and software developers.

## Final Words

One can use technology to teach the same old stuff in the same way. Integrated computer activities can increase achievement. Children who use practice software 10 minutes per day increase their scores on achievement tests. However,

> if the gadgets are computers, the same old teaching becomes incredibly more expensive and biased towards its dullest parts, namely the kind of rote learning in which measurable results can be obtained by treating the children like pigeons in a Skinner box....I believe with Dewey, Montessori, and Piaget that children learn by doing and by thinking about what they do. And so the fundamental ingredients of educational innovation must be better things to do and better ways to think about oneself doing these things. (Papert 1980)

We believe, with Papert, that computers can be a rich source of these ingredients. We believe that having children use computers in new ways—to solve problems, manipulate mathematical objects, create, draw, and write simple computer programs—can be a catalyst for positive school change.

## References


Biddlecomb, B.D. (1994). Theory-based development of computer microworlds. *Journal of Research in Childhood Education*, 8(2): 87–98.

Bowman, B.T., and Beyer, E.R. (1994). Thoughts on technology and early childhood education. In *Young children: Active learners in a technological age*, eds. J.L. Wright and D.D. Shade, 19–30. Washington, DC: National Association for the Education of Young Children.

Clements, D.H. (1991). Current technology and the early childhood curriculum. In *Yearbook in early childhood education, Volume 2: Issues in early childhood curriculum*, eds. B. Spodek and O.N. Saracho, 106–131. New York: Teachers College Press.

Clements, D.H., and Battista, M.T. (1992). *The development of a Logo-based elementary school geometry curriculum (Final Report).* NSF Grant No.: MDR–8651668. Buffalo, NY/Kent, OH: State University of New York at Buffalo/Kent State University.

Clements, D.H., and McMillen, S. (1996). Rethinking "concrete" manipulatives. *Teaching Children Mathematics*, 2(5): 270–279.

Clements, D.H., and Nastasi, B.K. (1992). Computers and early childhood education. In *Advances in school psychology: Preschool and early childhood treatment directions*, eds. M. Gettinger, S.N. Elliott, and T.R. Kratochwill, 187–246. Hillsdale, NJ: Lawrence Erlbaum Associates.

Clements, D.H., and Nastasi, B.K. (1993). Electronic media and early childhood education. In *Handbook of research on the education of young children*, ed. B. Spodek, 251–275. New York: Macmillan.

Clements, D.H., and Sarama, J. (1995). Design of a Logo environment for elementary geometry. *Journal of Mathematical Behavior*, 14: 381–398.

Cochran-Smith, M., Kahn, J., and Paris, C.L. (1988). When word processors come into the classroom. In *Writing with computers in the early grades*, eds. J.L. Hoot and S.B. Silvern, 43–74. New York: Teachers College Press.

Coley, R.J., Cradler, J., and Engel, P.K. (1997). *Computers and classrooms: The status of technology in U.S. schools*. Princeton, NJ: Educational Testing Service.

Confrey, J. (in press). Designing mathematics education: The role of new technologies. In *Education & technology: Reflections on a decade of experience in classrooms*, ed. C. Fisher San Francisco: Jossey-Bass and Apple Corp.

Davidson, J., and Wright, J.L. (1994). The potential of the microcomputer in the early childhood classroom. In *Young children: Active learners in a technological age*, eds. J.L. Wright and D.D. Shade, 77–91. Washington, DC: National Association for the Education of Young Children.

Emihovich, C., and Miller, G.E. (1988). Effects of Logo and CAI on black first graders' achievement, reflectivity, and self-esteem. *The Elementary School Journal*, 88: 473–487.

Escobedo, T.H., and Evans, S. (1997). *A comparison of child-tested early childhood education software with professional ratings*. Paper presented at the meeting of the American Educational Research Association, Chicago, IL, March 1997.

Forman, G. (1986). Computer graphics as a medium for enhancing reflective thinking in young children. In *Thinking*, eds. J. Bishop, J. Lochhead, and D.N. Perkins, 131–137. Hillsdale, NJ: Lawrence Erlbaum Associates.

Fuson, K.C. (1992). Research on whole number addition and subtraction. In *Handbook of research on mathematics teaching and learning*, ed. D.A. Grouws, 243–275. New York: Macmillan.

Genishi, C., McCollum, P., and Strand, E.B. (1985). Research currents: The interactional richness of children's computer use. *Language Arts*, 62(5): 526–532.

Gravemeijer, K.P.E. (1991). An instruction-theoretical reflection on the use of manipulatives. In *Realistic mathematics education in primary school*, ed. L. Streefland, 57–76. Utrecht, The Netherlands: Freudenthal Institute, Utrecht University.

Haugland, S.W. (1992). Effects of computer software on preschool children's developmental gains. *Journal of Computing in Childhood Education*, 3(1): 15–30.

Hennessy, S. (1995). Design of a computer-augmented curriculum for mechanics. *International Journal of Science Education*, 17(1): 75–92.

Hohmann, C. (1994). Staff development practices for integrating technology in early childhood education programs. In *Young children: Active learners in a technological age*, eds. J.L. Wright and D.D. Shade, 104. Washington, DC: National Association for the Education of Young Children.

Kamii, C. (1986). Place value: An explanation of its difficulty and educational implications for the primary grades. *Journal of Research in Childhood Education*, 1: 75–86.

Lemerise, T. (1993). Piaget, Vygotsky, & Logo. *The Computing Teacher*, 24–28.

Muller, A.A., and Perlmutter, M. (1985). Preschool children's problem-solving interactions at computers and jigsaw puzzles. *Journal of Applied Developmental Psychology*, 6: 173–186.

National Council of Teachers of Mathematics. (1989). *Curriculum and evaluation standards for school mathematics*. Reston, VA: Author.

Papert, S. (1980). Teaching children thinking: Teaching children to be mathematicians vs. teaching about mathematics. In *The computer in the school: Tutor, tool, tutee*, ed. R. Taylor, 161–196. New York: Teachers College Press.

Riel, M. (1994). Educational change in a technology-rich environment. *Journal of Research on Computing in Education*, 26(4): 452–474.

Ryan, A.W. (1993). The impact of teacher training on achievement effects of microcomputer use in elementary schools: A meta-analysis. In *Rethinking the roles of technology in education*, eds. N. Estes and M. Thomas, 770–772. Cambridge, MA: Massachusetts Institute of Technology.

Schery, T.K., and O'Connor, L.C. (1992). The effectiveness of school-based computer language intervention with severely handicapped children. *Language, Speech, and Hearing Services in Schools*, 23: 43–47.

Spodek, B., and Brown, P.C. (1993). Curriculum alternatives in early childhood education: A historical perspective. In *Handbook of research on the education of young children*, ed. B. Spodek, 91–104. New York: Macmillan.

Steffe, J., and Meinster, B. (1997). *Integrated computer activities to build science skills. (Df-Computer Application Series)*. Cincinnati, OH: South-Western Publishing.

Tan, L.E. (1985). Computers in pre-school education. *Early Child Development and Care*, 19: 319–336.

Thompson, P.W., and Thompson, A.G. (1990). *Salient aspects of experience with concrete manipulatives*. Mexico City: International Group for the Psychology of Mathematics Education.

Weir, S., Russell, S.J., and Valente, J.A. (1982). Logo: An approach to educating disabled children. *BYTE*, 7: 342–360.

Wilensky, U. (1991). Abstract mediations on the concrete and concrete implications for mathematics education. In *Constructionism*, eds. I. Harel and S. Papert, 193–199. Norwood, NJ: Ablex.

Wright, J.L. (1994). Listen to the children: Observing young children's discoveries with the microcomputer. In *Young children: Active learners in a technological age*, eds. J.L. Wright and D.D. Shade, 3–17. Washington, DC: National Association for the Education of Young Children.

Yelland, N. (1994). The strategies and interactions of young children in Logo tasks. *Journal of Computer Assisted Learning*, 10: 33–49.

# Science Assessment in Early Childhood Programs

*Edward Chittenden and Jacqueline Jones*

The momentum toward reform of science education brings pressures on schools and teachers to evaluate or otherwise account for children's progress in science. During an earlier era of neglect of science education, not much attention was paid to assessment and evaluation, but currently there is widespread interest at all levels of the educational system. This interest can bring with it a certain rush to judgment, but it also brings an opportunity to explore assessment alternatives that are fundamentally different from conventional evaluation methods.

Assessment can be defined as the process of identifying, collecting, and analyzing the records of learning in order to make informed judgments about students. Especially in early childhood, this process should support teachers' inquiry into children's learning more than identify discrete strengths and weaknesses. We know that learning takes time, young children need the chance to explore and make connections, and learning is social. Yet this very complexity of learning makes it difficult to see the "science" in children's activities. What does young children's science look like? How do you know it when you see it? Given this context, a first purpose of assessment in early childhood should be to enhance teachers' capacities to observe, document, and understand learning. Opportunities for thoughtful examination of children's learning may not be a routine part of the professional life of many teachers, but new approaches to assessment could provide occasions for such reflection.

*Edward Chittenden and Jacqueline Jones are with the Educational Testing Service in Princeton, NJ. Chittenden is a research psychologist, Jones a research scientist.*

## General Purposes of Assessment in Early Childhood

Educational assessments serve a variety of purposes and yield different kinds of results. The term *assessment* itself carries multiple meanings. Often, assessments are equated with testing. Statewide "assessment programs" of science are, in essence, statewide "testing programs." Sometimes the term suggests a more diagnostic function, as in the identification of children with special needs. At times, assessment invokes a wide array of procedures drawing upon various kinds of information, for example, classroom assessments of mathematics and early literacy that include the use of student work samples and portfolios.

A recent statement of principles and recommendations for early childhood assessment prepared by an advisory group for the National Education Goals Panel accentuates the importance of differentiating purposes of assessment (Shepard et al. 1997). (Distinctions of purpose are also prominent in the *National Science Education Standards* [National Research Council 1996].) As the panel's report indicates, the purposes determine the content of the assessment; the methods of collecting evidence; and the nature of the possible consequences for individual students, teachers, schools, or programs. In the past, serious misuse of tests and other instruments in early childhood has often stemmed from confusion of purpose. Instruments designed for one purpose, such as identification, may be completely inappropriate as instruments to measure the success of a program. With respect to early childhood education, four purposes provide the framework for the report's recommendations:

- assessments to support students' learning and development as part of instruction,
- assessments for the identification of special needs,
- assessments for evaluating programs and monitoring trends, and
- assessments for high-stakes accountability.

In this paper, we focus upon the first assessment purpose—to inform instruction and support learning. We start from the premise that the foremost function of classroom assessment in the early years is to enhance teachers' powers of observation and understanding of children's learning. We stress this function for two reasons: the rapid and variable nature of children's learning and the interactive nature of teaching. The classroom science envisioned in *Benchmarks for Science Literacy* (American Association for the

Advancement of Science 1993) calls for interactive instruction, which presumes that teachers can respond to young children's interests, background knowledge, and emerging skills. Whether the program is defined by science themes, units, or kits, the role of the teacher as observer and shaper of the classroom program is critical. Science instruction, which promotes children's inquiry and problem solving, must be guided by cues in the children's behaviors and language as well as by curriculum expectations.

## Guiding Principles of Preschool Assessment

### *Multiple Forms and Sources of Evidence*

Learning in early childhood is rapid, episodic, and marked by enormous variability. Even the most carefully designed assessment instrument cannot, by itself, capture the complexity of a child's understanding. Instead, evaluation of learning should be based on multiple forms of evidence from many sources. In active science programs, children make choices, voice opinions, and perform various investigations. In such settings, children might demonstrate their interests, understandings, and emerging skills through their conversations; their questions; their actions; and the work they produce, such as constructions, drawings, or writings. It is this sort of evidence that teachers can rely upon when evaluating whether an activity is meaningful and whether children are learning. The children's ongoing behaviors and their work are the *stuff* of teachers' everyday observations, records, and evaluations. In the case of science education, the richer the instructional environment, the broader the potential range of evidence for assessing learning (Bredekamp and Rosegrant 1995).

### *Evidence Collected Over Time*

Since young children's thinking reflects both developmental and experiential factors, teachers need to have a good sense of the appropriate instructional pace, allowing time for exploration and accommodation of new ideas. Children need time to revisit interesting phenomena; they need opportunities to ask the same question over and over again, perhaps in new or slightly different ways. Important ideas develop gradually—over days, months, and years—and are seldom the result of a single lesson or demonstration. Moreover, the development of thought is not neatly sequential, but rather marked by detours and explorations. Given this pattern of learning, indicators need to be collected on a regularly scheduled basis. For example, some portfolio assessment programs require that documents be collected at three or four specified periods of the year. Whatever the data collection method, the goal is to obtain records that reflect the child's developmental progress (Bredekamp and Rosegrant 1995).

### *Evidence Highlighting What the Individual Knows*

The evidence collected in early childhood assessments should go beyond the "deficit" model and highlight what children know. Teachers need to understand that children's "misconceptions" about natural phenomena are not necessarily unproductive; they may reflect keen observations and efforts to make sense of the world. For the teacher, this assessment requires an attitude of listening, of asking questions in an open way, and of attending to unanticipated answers. This stance toward assessment is exemplified when teachers collect information about children's interests and prior experiences as a step in planning instruction. For example, as an introduction to a unit on paper, a group of kindergarten teachers made experience charts from the things that children said were "made out of paper," "not made out of paper," or "not sure." The chart was revisited over the course of the unit.

### *Evidence of the Collective Knowledge of Groups of Learners*

Young children's science learning is inherently social. A teacher with whom we have worked remarked, "It's the many little conversations among children that really count for something" in promoting their ideas and observations. As an example, she described how a child discovered that by getting under the aquarium stand and looking through the glass bottom, one could witness a whole new dimension to the life of the fish tank, such as watching the sea worms tunneling in the sand. This experience caught on among the children, and, over the course of weeks, it promoted much talk and exchange of observations.

Although individual learning is typically the focus of classroom assessments, teachers need to be responsive to the patterns of interest and knowledge within the group. Documents reflecting the social dimension abound in young children's classrooms, such as displays of drawings, records of class discussions, and observations of group projects. Exploration into the understandings of a community of learners can provide insight into the prior knowledge and experiences that students bring to learning environments.

## Documentation as an Approach to Assessment

For a number of years, we have been meeting with teachers in elementary and preschool settings to explore classroom strategies for documenting children's science learning. Documentation is an approach to assessment that attempts to build directly upon evidence from teachers' everyday experiences of observing and listening to children and collecting samples of their work. As an approach, these methods are more open-ended than tests or checklists, yet more structured and systematic than incidental recordkeeping.

### *Children's Talk*

In our work to date, we have found that children's talk and language about natural phenomena is of particular interest to teachers and serves as a useful starting point. In the early grades, children's conversations and discussions constitute perhaps the single richest source of evidence to teachers concerning the substance of their students' ideas. However, in contrast to drawings, writings, and constructions, discussions leave behind no artifacts or documents for the teacher to review or consider. Children's talk is a facet of the teaching experience that tends to remain unrecorded and, hence, not ordinarily accessible to review.

### *Guidelines for Documenting Science Discussions*

In early education classrooms, most discussions and conversations among the children occur spontaneously and informally. However, there are also occasions when teachers bring the children together to share ideas and to talk about some activity. With some attention on the teacher's part, these occasions can become opportunities for investigating children's thinking.

The following guidelines were formulated with teachers who participated in a study of children's science learning. These guidelines were intended to facilitate the sort of discussions that are sustained by child-initiated questions and ideas and that allow children some control over the direction or drift of their remarks. In such settings, interactions among children may well bring out lines of thinking that are not so evident in individual interviews or group lessons, when children must deal more directly with the adult's agenda.

1. Discussions begin with open-ended questions, such as:
   "What are some things made of paper?"
   "Where have you seen shadows?"

"What do you know about water?"

"What have you noticed lately about our caterpillars?"

2. Children shape the agenda of the discussion.

The teacher sets the stage for conversation but does not dominate it. Children are allowed time and space to formulate ideas in their own terms and to pursue aspects of a topic that are of greatest interest to them. In general, teachers refrain from correcting or modifying children's comments.

3. Participation by all children is encouraged.

Teachers steer the discussion in a way that encourages comments from each child at some point. "Going around the circle" is one strategy. A discussion that involves most of the group will bring out evidence of the variety of interests among children; it will also highlight issues or questions that are commonly shared.

4. Records are made of each child's statements.

There are different ways of making a written record of the discussions. The fullest records can be made via tape recording or observer notes. Other records can take the form of "experience charts," lists, etc. Full transcripts are not needed, but an effort should be made to capture the key terms in any child's statement.

## Kindergarten Class Discussion: The Fish

The following discussion of a "dead" fish illustrates how the teacher provides structure for a conversation while allowing the children to shape the agenda. The result is that the matter of a fish's scales—the intended topic—becomes subsidiary to larger questions of whether the fish is dead and how to make that determination. In this particular example, the discussion led directly to a classroom investigation.

*Note: Donna Erickson, a Philadelphia kindergarten teacher, made this transcript.*

### *Fish Observation*

While I was reading *The Rainbow Fish* to the class, a child asked what "scales" were. A few days later I bought a fish at the supermarket and brought it to the class. Sitting in a circle, I showed the class how to feel the scales and invited them to tell us anything else that they noticed about the fish. Their discussion follows.

Darryl: "You got to scrape the scales off and then cook it."

Kate: "Was the other fish bigger in the book?" (Class thought the book fish was bigger.)

Blair: "I had a fish that die."

| | |
|---|---|
| Jennifer: | "Fish will swim in the water." |
| Sarah: | "I love to eat fish." |
| Ashley R.: | "I notice that he's dead." |
| Alina: | "It makes me remember trout fishing with my grandfather." |
| Derek: | "At my old school we fed fish in a pond." |
| Liam: | "The eye reminds me of jelly." |
| Frank: | "It reminds me of my alive fish. This one's alive. No. It's dead. I see the blood (around the eye)." |
| Earl: | "It's wet. I can feel its scales. I think it's alive." |
| John: | "I think it's dead." |
| Teacher: | "Why?" |
| John: | "I don't know." |
| Mickey: | "It smells bad. I think it's dead because I see blood." |
| Richard: | "It's dead." |
| Teacher: | "Why do you think so?" |
| Richard: | "Because fishes always die?" |
| Shelby: | "I like fish. I think it's alive." |
| Teacher: | "Why do you think it's alive?" |
| Donovan: | "I like fish." |
| Ashley H.: | "It feels like my cousin's fish. It's dead 'cause it ain't movin'." |
| Danielle: | "It's not movin'. It's dead." |
| Darryl: | Jumps up and yells. "No! Fishes swim in the water. You gotta put it in water!" (Many students agree.) |
| Zoe: | "It's dead." |

I got a plastic shoebox and filled it with water and put the fish in and set it before the children. I heard someone say, "It's sleepin'," and many agreed. I told the class that I'd put the fish on the table and they could keep their eyes on it. Kids went over throughout the day to check it out. Once there were screams of "It's moving! It's moving!" but then someone said, "No it's not. You just bumped the table and the water's movin'." By the end of the day when I asked the class about the fish, they all agreed that it was dead because it never moved.

## Lessons From Early Literacy Assessment

Over the past three decades, there have been major changes in assessment of early literacy, with some lessons for primary science. Where once *readiness* was narrowly measured, newer methods reflect a broader conception of literacy and recognize that children's steps toward reading and writing entail much more than alphabet recognition. These changes not only reflect theoretical advances but also extensive teacher participation in the observation of young children's efforts to make sense of print. Portfolios and other methods have played an important part in strengthening the teacher's capacities for inquiry and the teacher's contributions to new models of assessment. These methods have also demonstrated how assessments can build upon practice and how they need not interrupt teaching but can be embedded within instruction (Jones and Chittenden 1995).

Interest in science assessments brings the opportunity to explore methods that require a central role for early childhood science teachers. There are of course some critical differences between language arts and science instruction. For teachers, recognizing the science in children's behavior may well be more problematic than observing children's development as readers and writers, in part because of the teachers' own limitations of content knowledge. In addition, the boundaries of the child's development as a "scientist" are less clear. Children's ways of figuring out how the world works are not constrained by science lessons but cut across the curriculum areas. These points argue for greater involvement of teachers in the documentation and analysis of children's science learning, both for professional development and for the design of appropriate assessments.

## References and Bibliography

ILEA/Center for Language in Primary Education. (1988). *The primary language record forms* [Observations and Samples form]. Portsmouth, NH: Heinemann Educational Books, Inc.

American Association for the Advancement of Science. (1993). *Benchmarks for science literacy*. New York: Oxford University Press.

Bredekamp, S., and Rosegrant, T., eds. (1995). *Reaching potentials: Transforming early childhood curriculum and assessment.* (Vol. 2). Washington, DC: National Association for the Education of Young Children.

Jones, J., and Chittenden, E. (1995). *Teachers' perceptions of rating an early literacy portfolio.* Center for Performance Assessment Report MS# 95–01. Princeton, NJ: Educational Testing Service.

National Research Council. (1996). *National science education standards*. Washington, DC: National Academy Press.

Shepard, L., Kagan, S.L., and Wurtz, E. (1997). *Principles and recommendations for early childhood assessments*. Unpublished Draft. Washington, DC: National Education Goals Panel.

# Fostering High-Quality Programs

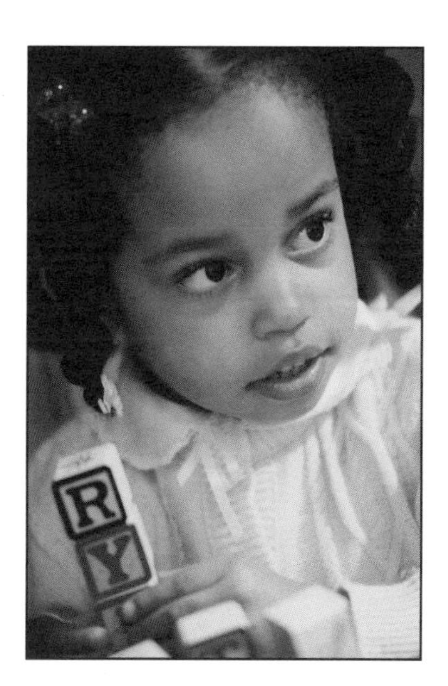

# Preparing Teachers of Young Learners: Professional Development of Early Childhood Teachers in Mathematics and Science

*Juanita Copley and Yolanda Padrón*

"I wish just once our principal would do some kind of workshop or in-service program just for us.... I am tired of being included in all the state testing workshops for grades 1–6!"

...A pre-kindergarten public school teacher

"I just want to teach three-year-olds at my preschool. I know I want to concentrate on their language and social skills. I really don't have to do any math or science, do I? I was never good at math and I always avoided science. In fact, that's one of the reasons I chose early childhood.... I don't have to know math or science!"

...A preschool teacher working on her associate's degree

"What is appropriate for young children? Is the water table the only science children should learn? Are counting activities with the calendar the only important things in kindergarten mathematics? I can't tell what is really going on every time I observe the kindergarten class. Early childhood classrooms are places I don't understand and I don't know how to help my beginning teachers or evaluate my experienced ones."

...A principal in a public elementary school

*Juanita Copley is an associate professor in the department of curriculum and instruction at the University of Houston. Yolanda Padrón is a professor in the department of curriculum and instruction at the University of Houston.*

> "We only learn about philosophy and theories. I am frightened to death of student teaching. I don't know if I can do it."
>
> ...A senior at a university one semester before student-teaching

> "I think I want to be a teacher...it sounds like a good job. More people should be working with young children and teaching them mathematics and science concepts. I want to get alternative certification and begin immediately...I figure I know all the mathematics and science I need...and teaching can't be that hard!"
>
> ...A chemist who had just enrolled in an early childhood curriculum class in a college of education

The professional development of early childhood teachers is of critical importance to the young children in our country. Enrollment in pre-primary education for ages 3 to 5 has increased 37 percent between 1984 and 1995, with the greatest increase in the four-year-old population. In addition, projections based on the most recent census indicate that there will be a 12 percent increase in total elementary school enrollment over the next 10 years. The testimonials cited, combined with the growing number of pre-primary programs, indicate a need for more pre-service training. It is also clear that both pre-service and in-service teachers will require specialized training to help them understand and instruct the increasing number of children with a variety of needs, children from various socioeconomic backgrounds, exceptional children, and children who are enrolled in state-required, pre-kindergarten programs.

Perhaps one of the greatest challenges, however, will be to address the educational needs of children who come from culturally and linguistically different backgrounds. Projections for the year 2020 indicate that people of color will comprise 46 percent of the student population. Not only will teachers continue to have many students from diverse cultural backgrounds, but many of the students will also have diverse language backgrounds. The number of school-age children from various language backgrounds has continued to increase at a rate of 12.6 percent; the overall student population has increased at a rate of only 1 percent. Most teachers have not received training in English as a second language or bilingual education, yet children who are learning English will be mainstreamed into "regular" classrooms.

Consequently, early childhood teachers will be expected to teach reading, writing, mathematics, and science as well as help children develop their oral English skills. To effectively teach these young children, both pre-service and in-service teachers require professional development opportunities in many areas. For the purposes of this paper, specific professional development opportunities for the early childhood teacher in the often-neglected areas of mathematics and science will be addressed.

According to the 1997 National Education Goals Report (National Education Goals Panel 1997), most mathematics teachers, while knowledgeable about reforms, do not exhibit many of the behaviors in the classroom suggested by those reforms. In addition, while 85 percent of in-service teachers participate in professional development programs, these same teachers receive little support at the beginning of their teaching careers through apprenticeship programs and other kinds of opportunities to interact with more experienced teachers. The Goals Panel report recommends that teachers' subject-matter knowledge and teaching skills in mathematics and science be strengthened. While little research addresses the competencies of early childhood teachers in mathematics and science, data collected from a four-year study revealed that early childhood teachers generally like to teach reading and other language-oriented skills. Mathematics or science or both were considered to be difficult subjects—ones they felt unable to teach.

Knowledge of and interest in mathematics and science, however, are not the only determinants of good instruction. Teachers serving economically disadvantaged, limited-English-proficient or lower-achieving students often devote less time and emphasis to the higher-level thinking skills so important to the learning of mathematics and science than do teachers serving more advantaged students. Since the expectations set for children to learn mathematics and science play an important role in determining students' achievement, the professional development of early childhood teachers must address the content, process, and dispositions associated with all learners in mathematics and science.

## The Current Status of Professional Development for Early Childhood Teachers

At present, there is a large distinction between childcare workers in daycare situations and certified teachers in public or private schools. Childcare workers have a wide range of credentials—some have attended one or two workshops, and others are practitioners with an associate or child development associate (CDA) credential. Certified teachers in public or private schools often have provisional early childhood certification and a bachelor's, master's, or doctoral degree.

Field experiences (short-term workshops, observations, simulations, practica, student teaching, and apprenticeship programs) are usually components of early childhood professional development programs. However, there is a wide range in both the quantity and quality of such programs. According to the 1997 National Educational Goals Report (National Education Goals Panel 1997), teachers in public elementary schools were more likely to participate in workshops or in-service programs (93 percent) than they were to take college courses (38 percent) or participate in activities sponsored by professional associations (50 percent).

Few professional development programs focus specifically on mathematics and science concepts in early childhood. Instead, the primary foci of professional development for early childhood teachers include definitions of developmentally appropriate curriculum, emergent literacy, management strategies, and the importance of play and strategies to improve social and emotional development.

The National Commission on Teaching and America's Future (1996) proposed five recommendations to improve and professionalize teaching. Two of those recommendations discuss the importance of professional development standards, specifically stating that teacher preparation and professional development programs should be reinvented and organized around standards. In addition, the Commission recommended the funding of new mentoring programs that adhere to these standards. Several other documents propose specific teaching standards for science and mathematics or for early childhood educators, but no document exists that integrates these standards.

## Integrated Professional Standards for Early Childhood Mathematics and Science Teachers

To create professional development standards for the childhood teacher with specific reference to mathematics and science, we synthesized three different sets of professional standards:

- *National Science Education Standards* (National Research Council 1996);
- *Professional Standards for Teaching Mathematics* (National Council of Teachers of Mathematics 1991); and
- *Guidelines for Preparation of Early Childhood Professionals: Associate, Baccalaureate, and Advanced Levels* (National Association for the Education of Young Children, Council for Exceptional Children, and National Board for Professional Teaching Standards, 1996).

The integration of these standards highlighted elements common to all of the documents *and* elements specific to early childhood or to mathematics and science. The documents indicate that both pre-service and in-service programs should provide the following experiences and opportunities for teachers, so they can meet these integrated professional standards.

### Integrated Standard No. 1: Develop Good Dispositions toward Mathematics and Science

Because of the frequently expressed phobic reactions of early childhood teachers to mathematics and science, this objective is an essential component of any professional development program for teachers of young children. Confidence, inspired by learning mathematics and science concepts, along with successful pedagogical experiences are critical to the development of early childhood professionals who can model positive dispositions toward mathematics and science for their young students.

### Integrated Standard No. 2: Experience Good Teaching in Mathematics and Science

No lecture, assigned readings, or observations can take the place of learning mathematics and science from an excellent teacher. The enthusiasm generated in the process of inquiry, the communication involved when solving problems with a small group, and the knowledge shared as different reasoning strategies are discussed are just a few of the processes that must be experienced as learners.

### Integrated Standard No. 3: Focus on Learning about Children and the Mathematics and Science Content of Specific Interest to Them

All early childhood teachers must be focused on children, how they learn, and their interests in the world around them. Listening to children's questions, assessing their understanding and unique abilities, understanding their natural inquisitiveness, managing and monitoring their learning, and designing their instruction to address children's learning are all skills that must be part of an effective professional development program.

### Integrated Standard No. 4: Participate in a Variety of Professional Development Opportunities Situated in a Learning Community

All three sets of professional standards emphasize the importance of apprenticeship and mentoring experiences shared with both experienced and inexperienced teachers. In addition, teacher-designed professional development experiences and opportunities that are varied and specific to identified needs are critical. The growing diversity of the children in our schools requires that early childhood teachers have opportunities to learn about children with special needs, children who are culturally and linguistically different, children who come from various socioeconomic backgrounds, and children who are enrolled in state-required, pre-kindergarten programs.

### *Integrated Standard No. 5: Demonstrate an Ability to Implement Integrative Curriculum*

An integrated approach to early childhood curriculum has been advocated and practiced for years in the early childhood community. This strategy has often resulted in an over-emphasis on those disciplines with which teachers feel most comfortable and a neglect of mathematics and science. The importance of intellectual integrity and sufficient time engaged in mathematics and science must be emphasized in a professional development program, along with practice in effectively integrating appropriate curriculum.

### *Integrated Standard No. 6: Utilize Appropriate Strategies to Establish Family Partnerships*

The importance of the family to the learning of the young child is emphasized in the National Education Goals that were established by Congress and the nation's governors in 1990. A parent's statements that "I was never good in math" or "I never had a scientific mind" are often communicated to young children and can stifle a child's desire to learn those subjects. Partnerships between school and family are essential to overcoming these and other obstacles in early childhood programs. If the objectives of the partnerships include the effective teaching of mathematical and science concepts, they can also lead to effective professional development for early childhood mathematics and science teachers.

## What Works?

Professional development programs that specifically focus on the early childhood teacher and mathematics and science instruction are difficult to identify. This section describes four of our programs that have focused on the professional development of early childhood teachers and their understanding of mathematics and science. All of these programs have taken place in the Houston, Texas, metropolitan area; education regions in Texas; or within the surrounding school districts or daycare centers. The following programs are in no way the only programs that work; however, there are several commonalities between these and other effective programs and with the integrated standards. The text focuses on a sequential description of each program as it was developed, the relationship between the program components and the integrated standards introduced in the previous section of this paper, and some general comments from participants in each program.

### *Trainer of Trainer Modules*

Eisenhower Grants and the Texas Education Agency funded the development of 37 professional development modules. As part of the grant, each of the regional centers in the state received manipulatives and software kits, along with the provisions to train trainers. Six modules, each consisting of a two-day interactive training session for in-service teachers, were written specifically for early childhood teachers in mathematics. The sessions combined learner-focused activities and pedagogy suggestions in a traditional professional development program. Activities were written and field-tested by classroom teachers and evaluated by early childhood experts and mathematics teachers. Highly dependent on the effectiveness of the trainer, this program addressed the Integrated Standards 1, 2, and 3: effectively developing good dispositions toward mathematics and science, experiencing good teaching, and focusing on children. It also provided a short-term professional development opportunity for in-service teachers.

### *Study Groups with Mathematics and Science Emphases*

Study groups have been used successfully in public schools, private schools, and daycare centers. Rather than a formal presentation by an expert, study groups are normally led by a teacher and based on reflections of assigned readings and classroom observations. In one local school district in Texas, representative early childhood teachers (pre-K to grade 2) meet monthly with other teachers in the district (grades 3 through 12) to discuss mathematical concepts and the *Professional Standards for Teaching Mathematics* (National Council of Teachers of Mathematics 1991). Representatives of this math council are responsible for sharing the information with other teachers at their school. Both teachers and administrators regard the study group as "a key professional development experience."

Another study group takes place in a highly respected daycare center in Houston. The teachers of three-, four-, and five-year-old children meet weekly to discuss readings specific to early childhood, child development, assessment, and mathematics and science. The lead teachers often consult early childhood experts to plan the group's activities or readings. Frequent evaluations of the program provide information for any needed adaptations.

The study-group approach has also been used in graduate classes. In one instance, eight doctoral students met weekly to discuss their videotaped interactions with young children. Students asked each other why they questioned children as they did, why they used certain materials, or why they responded to children as they did. They then analyzed and discussed their responses.

Each of these study groups has its own unique characteristics. All of them incorporated the Integrated Standards 1, 2, 3, and 4: the fourth involving teacher's participation in professional development opportunities. However, the teachers interacted only with other group members in their own learning community, reflecting and analyzing teaching experiences. Many of the discussions focused on integrated curriculum, more positive dispositions toward mathematics and science, and the assessment of young children.

The success of this approach is highly individual and specific to the group. One of the participants expressed it in this way: "I have an undergraduate and a graduate degree in early childhood and I have been to countless workshops on mathematics and science. This study group has changed my teaching more than any other type of professional development. The taping, critiques, discussions, readings, and sharing of our personal experiences has given me a new perspective on children and my interactions with them. I am a different teacher."

### *"Problem Solving for the Young Child": A Graduate Class*

To teach integrated curriculum appropriate for young children, we specifically designed a graduate class for early childhood teachers. The class deals primarily with assessing the young child's reasoning and thinking and the mathematics and science appropriate for young children. Since its introduction in 1994, the class has been taught four times to more than 100 students. At present, the Houston Independent School District uses the funds from its Eisenhower Grant to pay the full tuition of early childhood teachers from the district who are enrolled in the class. Class assignments include the following:

- classroom coaching experiences,
- presentation of a districtwide series of workshops on mathematics and science,
- mentorship of a pre-service early childhood teacher, and
- portfolios containing assessments of children's learning in mathematics and science.

This class has become a unique way to blend both pre-service and in-service professional development programs. During the class sessions, the in-service teachers experience good mathematics and science teaching, model the pedagogy necessary to transfer knowledge into practice, write and implement integrated curriculum, and then share their knowledge with pre-service teachers through workshops and mentoring programs. Adding the important early childhood standard of curriculum integration, this program (i.e., the class) successfully implemented the Integrated Standards 1, 2, 3, 4, and 5: the fifth involving the ability to implement integrative curriculum.

Student evaluations of this graduate class have been extremely favorable; in fact, four other school districts in the area have asked for similar programs. Teachers' enthusiasm for this type of professional development was reflected in comments such as:

- "The teachers presented the best workshop we've had in years!"
- "My mentor was fantastic...she really knew what she was talking about."
- "I learned so much from my assessment portfolio...four-year-olds are so smart...they solve problems in so many ways and come up with such interesting answers."
- "Math and science fit so well together....I can teach so much when I look at curriculum this way."

### *Collaborative Coaching Project*

To support the professional development of practicing early childhood teachers as well as new early childhood teachers, the University of Houston has sponsored a collaborative coaching community project for the past three years. An early childhood professor adopted a public elementary school close to the university. Every Tuesday, 18 pre-service teachers (teacher candidates who have not yet begun the field-based methods courses or student teaching) spend all day at the school in three collaborative coaching sessions, one teaching modeling session, and one debriefing session.

Working in teams of three, the pre-service teachers teach a previously planned mathematics or science lesson to kindergarten or pre-kindergarten students enrolled at the school. As required, the pre-service teachers use peer-coaching techniques with each other as they observe children and their peers' teaching. When the pre-service teachers are teaching, the regular classroom teachers meet with the university professor to discuss the mathematics and science lessons that they have taught during the week, experience some mathematics and science lessons as learners, discuss strategies for involving families in mathematics and science experiences, and reflect on the use of the professional teaching standards in their classroom. After each 70-minute session, the pre-service teachers discuss their teaching experiences with the university professor. Two more of these sessions occur during the day with other early childhood classes. In the remaining time, the beginning teachers and practicing teachers observe the university professor as she teaches a mathematics or science lesson in selected classrooms or as she coaches the practicing teachers.

The Collaborative Coaching Project covers all six of the integrated professional standards, with the emphasis on the sixth standard: establishing family partnerships. New and experienced teachers, children, and families are excited to teach and learn more mathematics and science. We received the following comments, which illustrate the positive disposition of all stakeholders.

- "I can't wait to teach math and science to young children....they've become my favorite subjects!"
- "I love it when the teachers from U of H come teach....we learned lots of neat stuff."
- "Math and science are everywhere at home...it's easy and fun to investigate with my child!"
- "I've learned more mathematics teaching these classes than I ever learned in four years of high school and four years of college!"

In addition, the assignments that required both beginning and experienced teachers to focus on children's learning and questions resulted in new knowledge for all members of the learning community. Increased scores on the state mathematics test and comments from teachers, including, "My children learned so much more when we did it this way," "I can't believe the ideas they have," and "Look at the way they solved this problem," underscore their belief in the program.

The Collaborative Coaching Project is indeed a program that benefits the pre-service teacher and the in-service teachers.

Pre-service teachers get needed experience in teaching pre-planned lessons to real children in real settings and then reflect on lessons taught with peer and professor input. In-service teachers experience professional development as part of their regular teaching day and then collaboratively reflect on lessons taught by themselves or the university professor. Most important, young children benefit from the project, which is the ultimate goal of any professional development. A child's comment as we left one day was simply expressed, "I hope you come back soon.... I'm so smart when you are here!"

## Conclusions

> "...The terms and circumstances of human existence can be expected to change radically during the next life span. Science, mathematics, and technology will be at the center of that change—causing it, shaping it, responding to it. Therefore, they will be essential to the education of today's children for tomorrow's world." (American Association for the Advancement of Science 1989)

In this paper, we have discussed the critical importance of professional development for early childhood teachers due to the growing and diverse student population in pre-primary education. In addition, we have indicated that, although knowledge and preference for mathematics and science are essential in the professional development of pre-service and in-service teachers, other considerations must also be addressed when designing staff development programs for early childhood teachers. For example, teachers' attitudes toward the development of higher-level thinking skills for economically disadvantaged, limited-English-proficient, or lower-achieving students must be addressed, since these skills are crucial to the learning of mathematics. The expectations set for children to learn mathematics and science also play an important role in determining students' achievement. The professional development of early childhood teachers, therefore, must address not only the content and processes involved in the teaching and learning of mathematics and science skills, it must also address dispositions associated with the teaching of concepts and skills in these content areas.

We have provided several examples of successful staff development programs specifically designed for early childhood teachers.

We created these programs to adhere to integrated professional standards for the early childhood teacher specifically in the areas of mathematics and science. The comments from teachers and students suggest that these programs were very successful. Furthermore, it is important to point out that the Collaborative Coaching Project has been very effective in terms of students' cognitive outcomes. Test scores for the state mathematics test have increased for those students whose teachers participated in the program. This increase in students' test scores is important, since the goal of staff development is to help teachers improve their teaching and, ultimately, to improve students' academic achievement.

## References

American Association for the Advancement of Science. (1989). *Science for All Americans*. Washington, DC.

National Association for the Education of Young Children, Council for Exceptional Children, and National Board for Professional Teaching Standards. (1996). *Guidelines for preparation of early childhood professionals: Associate, baccalaureate, and advanced levels*. Washington, DC: National Association for the Education of Young Children.

National Commission on Teaching and America's Future. (1996). *What matters most: Teaching for America's future*. New York: Author.

National Council of Teachers of Mathematics. (1991). *Professional standards for teaching mathematics*. Reston, VA: Author.

National Education Goals Panel. (1997). *National education goals report*. Washington, DC: Author.

National Research Council. (1996). *National science education standards*. Washington, DC: National Academy Press.

## Other Readings

Guskey, T.R., and Huberman, M., eds. (1995). *Professional development in education: New paradigms and practices*. New York: Teachers College Press.

Saracho, O.N. (1993). Preparing teachers for early childhood programs in the United States. In *Handbook of research on the education of young children*, ed. B. Spodek. New York: Macmillan Publishing Company.

Spodek, B., and Saracho, O.N., eds. (1990). *Early childhood teacher preparation: Yearbook in early childhood education, Volume 1*. New York: Teachers College Press.

## Web Sites

National Education Goals Panel: http://www.negp.gov/

National Center for Education Statistics: http://www.nces.ed.gov/

Federal Interagency Council on Statistical Policy: http://www.fedstats.gov/

National Association for the Education of Young Children: http://www.naeyc.org/

# Partnerships Among Families, Early Childhood Educators, and Communities to Promote Early Learning in Science, Mathematics, and Technology

*Heather B. Weiss*

> I don't think our kids have what they need at the school. My granddaughter wants a computer; she wants to do a computer so bad.... I can never in a million years get this little girl a computer.... The schools here don't supply this for the little ones.... That's why I'm looking for a school where she will have computers every day at school.
>
> (A very poor, urban grandmother raising her six-year-old granddaughter explains her plans to move in the next year.)

As this grandmother's comment suggests, now is a very opportune time to launch a serious, major effort to forge partnerships among families, communities, and the full range of early childhood educators. These partnerships should focus more attention and resources on science, mathematics, and technology education for children under five years of age. Parents know that their children need knowledge and skills in these areas to succeed in the 21st century. This paper will

- discuss why now is the right time for educators to partner with parents and others;
- briefly assess what research and practice in parent education and in family support indicate about how to successfully involve parents and communities; and
- suggest key components for a national, community-based strategy.

This strategy contains several key points. First, the goals and practices of developmentally appropriate early childhood education arguably overlap with the goals and processes of early science and math education. This will greatly facilitate partnerships.

Second, merely giving parents materials on how to teach children about science does not educate parents or involve the community. A materials-only approach is particularly ineffec-

*Heather B. Weiss is director of the Harvard Family Research Project.*

tive with those families most in need of help (perhaps because they are less educated, geographically isolated, or poor), and it does not work that well with other parents, either. Personal interaction with families, special training for them, and other supports are also needed.

Third, the assumption that parents want to help their children and can do so must undergird any effort. Parents must be involved in order to create a viable strategy.

Fourth, it is important to take advantage of and build on existing systems for delivering both early childhood and science, mathematics, and technology (SMT) education, rather than create new or single-institution, isolated delivery systems. All communities already have agencies and institutions, sometimes linked to each other, that can or do promote SMT education. In the interest of developing a community-based strategy, these organizations could be encouraged to collaborate more with one another, with parents, and with other local providers of care and education for young children.

Fifth, a community strategy that involves parents and multiple agencies (schools, libraries, museums, science and nature centers, childcare centers, family resource centers, health clinics, etc.) is more likely to reach most young children in an effective, affordable, and sustainable way. Such a strategy is ambitious, requiring new collaborations and some additional resources at the community level. It also requires formative evaluation to develop and implement. Provisions should be made to assess the efficacy, cost-effectiveness, and sustainability of a few community-based efforts such as those recommended in this paper before these efforts are expanded.

Sixth, attention must be paid to creating the national (and perhaps state) infrastructure necessary to support community-based efforts. This strategy includes building relationships among the various professional communities that have something to offer early childhood education: early childhood educators; educators in science, mathematics, and technology; family support professionals; and other professional groups whose community institutions bear on early childhood education (libraries, museums, science centers, the media, etc.). Also essential are a clearinghouse for innovative materials and approaches and for information about effective models and practices; grants to stimulate community-based work and collaboration; and provision for the ongoing evaluation of any strategy that is implemented.

## Current Opportunities and Challenges

Increasingly, policymakers and the general public are coming to understand the critical importance of children's early development to their later learning. The president, Congress, governors, and state legislators are now willing to invest in new and enriched early childcare and education. The time is right to create new partnerships among parents, early childhood educators, SMT educators, and related community groups and institutions that aim to support early childhood development. The needs and interests of these various groups are converging due to a number of factors, including the following:

- broad public and parental recognition that early brain development requires a stimulating and nurturing environment and personal interactions;
- the country's need for high-quality care for all children, including those living in families moving from welfare to self-sufficiency;
- the need of early childhood educators and schools to promote the readiness of young children, particularly given increasing public pressure to hold schools accountable for educational outcomes;
- educators' recognition that interest and skills in science, mathematics, and technology need to be fostered early in childhood and then expanded in the K–12 period; and
- widespread recognition in both the public and private sectors of the need for a scientifically, mathematically, and technologically skilled workforce for the 21st century.

The likely result is more federal, state, and local support for early childhood services and a growing system of services to foster young children's interest and skills in science, mathematics, and technology.

Establishing new federal and state early childhood programs and expanding the existing programs—for example, Georgia's new universal Pre-K program and North Carolina's Smart Start—also increase the number of stable systems that deliver early childhood education and the number of organized and ready partners for SMT education. Partnerships between those who care for young children and SMT educators are also becoming more feasible because the two groups share many goals and views about what constitutes developmentally appropriate practices. To wit, each is committed to creating learning environments that foster young children who are curious, eager, questioning, and intent on finding out about the world through observations and experiments. Partnerships between early childhood educators and SMT educators seem most likely to develop because the emphasis on SMT is integral to the definition of high-quality early childhood education. Working together, early childhood educators and SMT educators could substantially enrich the learning environment (including the community) for young children.

Because there is not, nor is there likely to be, a single, universal delivery system for early childhood services (akin, for example, to the public school), one could argue that it will be very difficult to reach all or most young children with early education that emphasizes science, mathematics, and technology. The lack of a single system, however, arguably provides an opportunity to create a rich array of supports through partnerships within a community. These partnerships can create a more intense, varied, and diverse set of learning opportunities for young children. Most communities now have some sort of formal or informal early childhood coordinating group. The challenge for national, state, and local professional groups interested in early childhood and SMT education is to identify and work with the coordinating group and get its help in enlisting other relevant groups in a communitywide effort. Some of the agencies, institutions, and groups that can be tapped to create a more comprehensive strategy for promoting SMT education within a community include:

- childcare centers;
- family daycare networks;
- childcare information and referral agencies;
- kindergarten and pre-kindergarten programs;
- Head Start;
- parent organizations;
- family resource centers and family support programs (including home visit programs);
- service centers, nature centers, and museums;
- libraries;
- health clinics and public health agencies;
- schools;
- cable and public television stations;
- voluntary groups;
- electronic sites on the World Wide Web; and
- newspapers and other media.

These types of community agencies and groups create a learning environment for young children. Their involvement, or lack thereof, will influence the intensity, creativity, reach, comprehensiveness, and effectiveness of any SMT education effort. In a community-based effort, such groups and institutions *can* come together to create SMT education programs and to share materials and resources.

## Lessons from Community-Based Initiatives for Parent Education and Family Support

Research shows that young children learn more from opportunities to experience, shape, and understand their environment than they do from their efforts to master discrete facts or skills. In addition, a child's disposition for learning develops through interactions with his or her parents, family members, and other caregivers and through experiences such as storytime at the library, trips to the zoo or museum, a show on The Discovery Channel, and the like. Children's early learning, including the precursors for learning science, mathematics, and technology such as curiosity and observation, occurs through a dynamic process involving child, family, and community influences. This research suggests that SMT education efforts should include all the key players in children's lives, including parents, providers of early childcare and education, and relevant community organizations.

Now that developmental research has substantiated the critical role that early parenting and community supports play in children's early development and later readiness to learn at school, parents are struggling to balance their efforts to be good parents with their work and the other demands on their time. There is much evidence that many families, constrained by busy schedules or tight budgets or both, are under a great deal of stress. The same can be said of the web of community institutions that work with families and young children: Time and money are often scarce, even as demand for services is increasing. Many communities struggle, therefore, with the question of how to create systems of early childcare and education that recognize the importance of parent and community involvement *and* the constraints on would-be participants. This overall challenge is critical to consider in building parent and community partnerships to support young children. It is also critical to understand that any successful effort will require some new and carefully targeted resources.

Past efforts to work with the parents of young children provide many important lessons. Many of today's stable and effective programs for involving parents in early childhood education were created after earlier efforts had failed, for example, programs that no parents ever joined or the ones that they quickly deserted. Many of these failures were didactic programs that did not offer parents what they wanted when they could use it. Years of research on implementation and practice-based experience suggest the following lessons.

- Include parents and relevant institutions or groups from the start when designing the program.
- Programs that draw on parents' strengths and empower them to do more for their children are more likely to attract and retain parents than those that do not. These programs underscore and reinforce the importance of the parents' role in child development. Successful programs also educate parents on *specific* ways to promote their child's learning.
- Parenting activities and materials to promote children's early learning must be free or low cost and fit into everyday life.
- Parenting activities and programs need to focus on developmentally appropriate practice (as opposed to, say, "skill and drill"). They need to convey specific ways that parents can help children of a particular age to learn and discover. All those who work with young children should reinforce this approach in their work with parents and other caregivers.
- Effective programs help parents and other caregivers access the array of community institutions that support early learning and development. They are creative in building partnerships and collaborations with these institutions.
- Effective programs create partnerships with parents and with other agencies, partnerships that allow *all* these groups to share their expertise.
- In general, communities that create diverse and rich opportunities for early learning have had some formal or informal early childhood coordinating group that helps connect the players. Such groups are designed to build a comprehensive and high-quality system of early childhood services. They promote collective training and funding opportunities, identify gaps in services, and determine how to use new resources most effectively.
- While much can be accomplished through better coordination of existing resources and voluntary efforts, major initiatives require additional resources—time and money—to support their work and to provide an incentive to innovate.
- Effective early childhood efforts build links to the K–12 schools in order to prepare children for school entry, share expectations and expertise, and promote continuous parent involvement.
- Effective programs recognize that the *whole* community has a role to play in promoting early learning. They

therefore help to develop communitywide strategies that engage many institutions. This approach, in turn, provides a rich and diverse array of learning resources and activities that have the collective potential to reach the vast majority of young children.

## Key Components of a National Strategy to Support Community-Based Early SMT Learning

This paper has argued that now is an opportune time to launch a major effort to forge partnerships among families, communities, and other concerned groups that will focus more attention and resources on science, mathematics, and technology learning in the early years of childhood. To stimulate debate about what such an effort might look like, some of the key design features of a major initiative are outlined in the following text:

- a combined national and local approach in an effort to maximize efficiency and individualized approaches by communities.
- an emphasis on beginning with the customer, that is, initiating outreach to parents and other concerned parties when designing both the national and local efforts.
- an effort to involve the many players necessary to develop a comprehensive and intensive initiative at both the national and local levels.
- an effort to map all available resources and possible players and to foster new and creative partnerships (for example, among networks of family childcare providers, science centers, and kindergarten and elementary school teachers). Information on existing innovative approaches should be disseminated at the beginning of this process and should continue throughout its course.
- an emphasis on formative evaluation and continuous learning by partners as well as on an effort to examine whether the overall strategy and the community-based iterations are successful in achieving their goals and outcomes for children.

## References and Bibliography

Dunst, C. (1994). *Key characteristics and features of community-based family support programs.* Commissioned Paper. Chicago, IL: Family Resource Coalition.

Galinsky, E., and Weissbourd, B. (1992). Family-centered child care. *Yearbook in early childhood education: Issues in child care 3,* eds. B. Spodek and O.N. Sarcho. New York: Teachers College Press.

Lopez, M.E., and Hochberg, M. (1993). *Paths to school readiness.* Cambridge, MA: Harvard Family Research Project.

Powell, D. (1991). Early childhood as a pioneer in parent involvement and support. In *The care and education of America's young children: Obstacles and opportunities,* ed. S. L. Kagan, 91–109. Chicago, IL: University of Chicago Press.

Schultz, T., Lopez, M. E., and Hochberg, M. (1996). *Early childhood reform in seven communities: Front-line practice, agency management, and public policy.* Washington, DC: Office of Educational Research and Improvement, U.S. Department of Education.

Weiss, H., and Harvard Family Research Project. (1993). *Building villages to raise our children.* Cambridge, MA: Harvard Family Research Project.

# Playing Fair and Square: Issues of Equity in Preschool Mathematics, Science, and Technology

*Rebecca S. New*

The period of early childhood is a time in which children's development is especially ripe for the enhancement of numerous social, emotional, and cognitive capacities. Contemporary research also confirms that the experiences of three-, four-, and five-year-olds are significant precursors to children's subsequent learning and school achievement. Unfortunately, the young child's readiness to learn also includes, by definition, a vulnerability to a lack of certain educational experiences. While some of the variation in children's learning and development is the result of purposeful choices made by parents, teachers, and other adult members of their communities, other differences result from lack of opportunity, motivation, understanding, or some combination of factors. Research on preschool children's knowledge, skills, and dispositions in math, science, and technology has consistently demonstrated differences in children's learning as a function of gender, economic and socio-cultural factors, and developmental characteristics. The set of differences associated with educational inequities serves as the focus for this paper.

While early childhood educators have questioned the appropriateness of "too-early" instruction, recent research and classroom practice validate the premise that educational opportunities associated with mathematics, science, and technology are not only highly feasible but, if done right, can contribute to children's learning and development in other areas as well. Thus, for many in the field, the debate has shifted from a question of whether or when to instruct young children to one of how. Issues of equity must assume a position of prominence as educators consider how

*Rebecca S. New is associate professor of education at the University of New Hampshire.*

to encourage each young child's emerging mathematical understandings, scientific thinking and problem-solving skills, and potential technological literacies. This paper will address two questions associated with the challenge of equity.

1. What are the key issues of equity with respect to math, science, and technological education in the preschool-age period?
2. How can more effective teaching contribute to greater equity, not only in those specific domains, but within the classroom and the larger society as well?

These questions warrant more complex responses than can be fully provided in this paper. The following discussion focuses on two broad themes: educational equity in an inequitable society and equity in inclusive early childhood classrooms. These themes illustrate both the problems encountered when addressing the issues of equity in math, science, and technology as well as the potentials inherent in some early childhood programs.

## Placing Educational Inequities in Context

Children's development and learning are influenced and interpreted by the larger socio-cultural context. Comparative studies support the premise of cultural diversity in beliefs regarding children's needs and abilities and interpretations of appropriate educational experiences for optimal development (cf., studies on Japanese child care and early education by Lewis 1995). Beatty's recent (1995) analysis of the history of preschool education in the United States joins research on other nations (Woodhead 1996). This research confirms the presence of diversity in (1) perceptions of high-quality early childhood programs across cultures and (2) access to high-quality early childhood programs within cultures. The notion of "diversity as adversity" is particularly relevant as it pertains to unequal and inequitable learning opportunities for young children in contemporary American society (New and Mallory 1996).

Many believe that the purpose of American educational institutions is to "follow, reflect, and reproduce the nature of the society in which they exist" (Oakes 1985, page 200). Thus, U.S. schools and their curricula have historically promoted autonomy and individual competence, an educational agenda that has placed some children at significantly greater disadvantage than others. In spite of numerous national initiatives over the past three decades that have targeted diverse populations for more equitable treatment, the contrast between some children's educational opportunities and those available to other children remains stark. Throughout the 20th century, discrepancies in young children's educational experiences have been documented as a function of their membership in racially, culturally, and linguistically diverse populations, with still other differences associated with gender and developmental diversity. Among the most glaring of such discrepancies is children's unequal access to high-quality educational programs in the preschool-age period.

The United States is unique among industrialized nations in its failure to systematically provide some form of educational opportunity for all children three- to five-years of age. Adhering to the view that the responsibility for the very young child is familial and private rather than social and public, the periodic investments at the local, state, and national levels have been incapable of responding to the expressed need for affordable and high-quality early care and education. For example, since inception, such educational services as Head Start have been remarkably underfunded, typically serving less than one-third of the children who are eligible. The shortage of high-quality programs is just part of the problem, however. Issues of inequitable access to government-funded programs are joined by an overall decline in the perceived quality of numerous private and community-based programs over the past two decades, in spite of a growing knowledge base regarding the characteristics of high-quality preschool programs (Kagan and Cohen 1996).

Joining these claims of inadequate coverage and deteriorating program quality is the possibility that the nature of the field's targeted programs may "exacerbate the very problems they were designed to ameliorate" (New and Mallory 1996, page 150). Early intervention programs are designed for young children with special needs, while other categorical programs such as Head Start are limited to the most impoverished families. The mandated deficit interpretations of eligible children or their families result in segregated programs that often preclude exposure to and experiences with classmates of varying needs and abilities. Furthermore, the curriculum in such programs typically emphasizes certain aspects of development—such as physical, social, and emotional development—at the expense of other developmental areas, for example, cognitive or preacademic gains. In spite of the overall high quality of many of these targeted programs, such characteristics may limit program effectiveness in achieving educational parity.

Children entering preschool also bring with them evidence of social and educational inequities. Even those children fortunate enough to attend a high-quality preschool or kindergarten program will still demonstrate diversity in their readiness to learn particular skills and concepts, reflecting the prejudices and the potentials of the larger society. Some children will have had numerous opportunities to visit science museums, play with tanagrams on the living room floor, and experiment with the technological mouse attached to their family's computer. They will have acquired a vocabulary for discussing their ideas and experiences in the domains of mathematics, science, and technology. They may also have learned a great deal about the role of and value assigned to such knowledge in the larger adult society.

Other children enter into early childhood settings in the hope that they, too, will learn the skills and acquire the concepts deemed necessary for productive and meaningful participation in the larger world. Some of these children, however, will have had little or no exposure to the tools or the talk of mathematical or scientific endeavors. Their insistence that a mouse is an inhabitant of their family's basement will be a source of amusement to other children "in the know." Their lack of familiarity with contemporary technological tools and discourse may or may not lead to appropriate educational opportunities, depending on a number of key factors that influence what happens inside the classroom.

## Inequities on the Inside

Professionals in early childhood education began addressing issues of racism and sexism in the teaching of young children long before the war on poverty and the multicultural education movement (cf., Dewey 1911; Goodman 1952). And yet, when children arrive at public school settings (kindergarten and beyond), those with a history of early intervention services are often assigned to readiness classes; others are grouped based on so-called risk indicators, including socio-cultural or economic characteristics (Oakes 1985). Tracking of this sort perpetuates the class and racial inequalities of American society. It also widens the divide between children excluded from participating in other models of education, including gifted and talented programs—where an emphasis on mathematical knowledge, scientific endeavors, and technological literacy is almost guaranteed—and those invited to participate. These inequalities in the resources and programs available in the preschool and subsequent elementary-age period increase the likelihood that official bodies

of high-status knowledge and ways of thinking remain the property of select groups of children and their families. It is essential to place discussions of equity in math, science, and technology within this larger social, political, and economic context.

In response to this structural inequity in the public schools, the last two decades have witnessed a steady increase in the number of publications that assist teachers in responding more equitably to diverse populations of young children (cf., Derman-Sparks and the A.B.C. Task Force 1989; Kendell 1996). Efforts to create more inclusive educational programs for children with developmental differences and other special needs have also intensified; some of these children also represent racial and linguistic minorities (Harry 1992; Mallory 1998). In short, early education professionals have much on which to pride themselves. However, the field of early childhood education has not been immune to the dilemmas associated with efforts to respond more equitably to children who are diverse as a function of gender, development, or cultural background. There are also key features of the field of early education itself that may inadvertently contribute to inequities in children's learning in the areas of math, science, and technology. These features include teacher attitudes regarding diversity; teachers' personal and professional knowledge of math, science, and technology; and teacher beliefs about how children learn. Establishing fair, feasible, and relevant educational goals for diverse populations of young children remains a central—and controversial—challenge for early childhood educators.

## From Deficits to Deference

Adult images of children have historically defined the parameters and prerogatives of child care and early education. As a result, teacher interpretations of the meaning of differences among children directly influence curriculum goals and strategies. Teacher beliefs about the mutability of such differences also influence their responses to children. Such beliefs might prompt, for example, lesser expectations for girls to participate in scientific problem-solving. Even the kinds of questions teachers ask vary as a function of teacher expectations of competence, as when, for example, boys are more often called upon for complex explanations of mechanical and conceptual concerns while girls are asked to share "facts." Furthermore, some differences are viewed as legitimate expressions of children's diverse interests and learning styles rather than as indicators for curriculum planning. For example, teachers might use the topic of science and technology primarily as a means of attracting the attention of otherwise disengaged learners or as an occasion to single out children with prior knowledge as "class experts," rather than as a means to promote the knowledge and literacies of all children.

Throughout the last several decades, a "different strokes for different folks" philosophy has prevailed, which supports educational practices that respond to children's individual differences and family lifestyles. It is difficult to find fault with a pedagogy that is grounded in knowledge about and respect for children and their families. If implemented uncritically, however, this sensitivity to differences can further exacerbate inequities in children's learning. At the least, it can interfere with another principle central to an education for a democratic, pluralistic society: that all children are entitled to gain access to the skills and knowledge regarded as social capital in the dominant culture (Delpit 1995). This "different strokes" mentality can also undermine the role and the responsibility of teachers in changing rather than deferring to patterns of work, play, and social behavior that are less advantageous to some children.

In fact, a deference strategy often characterizes early childhood programs that adhere to a multicultural philosophy of learning about and responding to children's family backgrounds and individual learning styles and abilities. Even as teachers strive to create a more inclusive educational environment that reflects the lives and lifestyles of all the children in the class rather than just a few, teachers' deference to student differences may actually lead to an implicit acceptance of disappointments in educational outcomes. The complexity of this issue cannot be overstated. Put simply, in their commitment to the multicultural, anti-sexist, and inclusive education movements, early childhood educators may have miscalculated the effect on school curriculum and children's learning when they emphasize the value and legitimacy of children's differences more than children's need for essential common skills and understandings.

Non-critical acceptance of expected variation in children's interests or abilities does little to modify that variation. Deference to student differences may also influence some early childhood teachers to respond more systematically to children's social and behavioral developmental needs than to their intellectual ones. For example, early childhood special education services frequently defer emphasis on intellectual content or academic goals in favor of self-help skills and social relations. Such variations in classroom practices as a function of gender, disability, or family background are similar in intent and outcome to many of those practices described in the previous discussion, in that children's perceived differences require differential educational responses. Despite the best of intentions, however, when such deference to children's individual and cultural differences encourages neglect of essential educational goals, such responses increase rather than eliminate issues of inequity.

Recent studies on the role of computer technology in the early childhood classroom can illuminate this problem of teacher deference to children's differences. Initially viewed as an ideal response to the gifted child's natural curiosity and interest, computers in many early childhood classrooms continue to be reserved for special children or as a special privilege. Children with identified learning disabilities may have a mandated provision for computer-assisted instruction, while computer time for typically developing children serves as a reward for good behavior. This interpretation of the computer's role in the classroom serves as an especially powerful motivator for those children who may already have considerable computer experience outside of the classroom. Thus, those children who are knowledgeable about computers get additional opportunities to improve upon existing competencies. Those children who need more purposeful learning experiences coupled with teacher support and guidance continue to lag behind their peers.

## Teacher Attitudes and Knowledge

Teacher attitudes and knowledge may also account for much of the inequitable treatment of preschool mathematics, science, and technology. The field of early childhood education has struggled for much of the second half of this century to establish a reputation of professionalism. However, the knowledge base deemed essential for teachers' scientific and professional status derives almost exclusively from the child study movement and the field of developmental psychology. Few states require early childhood educators to have formal professional knowledge in the content areas as a condition of certification. Consequently, the experiences in science, mathematics, and technology that many early childhood educators bring with them to the classroom are limited by their personal histories as learners in those domains. Thus, children become accustomed to a female teacher's comments that "boys...know more about how that thing works than I do." Such teachers are more likely to use computers as crutches rather than as a valued educational tool. Some female teachers' willingness to display their own lack of knowledge in computer technology also reinforces the gender stereotype that computers are not essential for girls' development.

Teacher attitudes about specific subject matter also influence their approaches to issues of equity. For many in the field of early childhood education, experiences in math, science, and technology are generally regarded as less critical to children's development than are play-based experiences. It is also the case that a vast majority of early childhood educators are women, whose anxieties in certain learning situations are now the topic of study in the newly defined domain of "hot cognition." This conceptualization of the interface between teachers' emotional anxiety, social supports, and intellectual competence has contributed substantially to our understanding of both the causes of and potential solutions to poor academic performance in mathematics, science, technology, and other areas. By explaining the relationship among affect, personal relevance, and intellectual activity, hot-cognition theory also helps to explain teacher reluctance to engage in explorations—whether mathematical, scientific, or technological—about which they feel little competence or confidence. Indeed, many early childhood educators readily admit their reticence to explicitly incorporate science and technology into their curriculum, based on their own incomplete understandings of these academic domains.

## Beliefs About How Children Learn

For much of the second half of this century, early childhood professionals have debated the role of instruction in children's learning and development. Based in great part on Piagetian interpretations of the child's capacity to construct knowledge out of concrete experiences with objects of the material world, much of the early childhood literature has emphasized the value of play in a child-initiated curriculum. This interpretation of children's developmental needs, in turn, has contributed to a view of the teacher's role that has often been limited to preparing the physical environment and then following the child's lead, rather than imposing pre-determined educational goals. This position was supported in the form of guidelines for "developmentally appropriate practice" (Bredekamp 1987). These guidelines were published by the nation's largest early childhood professional organization (NAEYC) in response to increasing pressure from elementary school teachers, administrators, and some parents to start formal academic instruction in the preschool-age period.

The concept of developmentally appropriate practice and the associated guidelines played a valuable role in drawing educators' and parents' attention to the knowledge base of child development and especially to the role of play in children's social and cognitive development. Based on the premise of age and individual differences as determinants of appropriate practice, this concept has also been used to support teachers' willingness to accept children's choices, even when such choices reinforce gender-based or cultural differences in academic competencies. Thus, for example, some early childhood teachers hesitate to interfere when girls gravitate to the dramatic play area even as the boys lay claim to the blocks, or when the non-English-speaking child prefers solitary play with puzzles over the more verbal and scientific activity associated with the water table.

The role that teachers often assign themselves with respect to children's learning of mathematics, science, and technology also reflects their views about how children learn (Fennema et al.1993). Although the concepts of play and teacher planning of educational experiences are not necessarily in opposition, the notion that conceptual understandings are best pursued by children through play and other child-initiated activities has frequently served to eliminate the need for purposeful teacher planning in domains such as mathematics and science.

A recent article, "If We Call It Science, Then Can We Let Them Play?" (Goldhaber 1994), clearly articulates the relationship between constructive play activities and important scientific constructs. The title also reveals the tension felt by many teachers when attempting to respond appropriately to children's developmental needs and relate those needs to academic goals. While the "hands-on" maxim provides children with valuable opportunities to manipulate and explore the characteristics of scientific materials and mathematical concepts, teacher hesitancy to provide more systematic opportunities for children to reflect upon their ideas and their work makes it less likely that such play-based experiences will guarantee significant conceptual change. This minimization of the teacher's role is supported by the belief that children learn at their own pace, when, in fact, sometimes it is the adults who are moving slowly.

## What's the Good News?

Up to this point, the discussion has focused on the more problematic aspects of achieving equity in classrooms where teachers struggle to respond appropriately to the diverse needs, interests, and capabilities of children; to confront subjects about which they feel little personal or professional commitment; and to balance developmental goals with academic expectations. The final section of this paper considers the challenge of equity from a more optimistic point of view. It is based on recent advances in our understandings of how children learn and a reconceptualization of developmentally appropriate practices in the early childhood curriculum.

## Learning as a Social Process

Researchers in anthropology, psychology, and education have expanded prior conceptions of the child's solitary construction of knowledge to emphasize the role of the socio-cultural environment in children's learning. Contemporary theory on child development highlights the relational processes by which children and adults alike acquire the knowledge, skills, and attitudes deemed normative and desirable within particular socio-cultural contexts. Summarized most often as a theory of social constructivism, this perspective regards learning as both a social and cognitive process dependent upon interpersonal exchanges and upon optimally challenging tasks to complete and ideas to contemplate (Berk and Winsler 1995).

This new theoretical paradigm supports the premises laid out earlier in this discussion, primarily that children's knowledge of math, science, and technology—like any aspect of children's learning—is informed, influenced, and judged by the socio-cultural contexts and social exchanges that characterize their lives. Even very young children learn what is important, tolerated, and expected as they observe and participate in early educational experiences. Thus, gender-role stereotypes, ethnic identity, and self-image as a learner are among those understandings that develop during the period of early childhood (New 1998a). However, research also suggests that young children have the cognitive capacity to understand the difference between what people can do and what they usually do (Meece 1987). Such studies are essential to supporting teacher efforts to promote more equitable learning opportunities for all children, regardless of gender or ethnic identity.

These theoretical premises have significant implications for the role of early schooling in the formation of skills and knowledge—as well as attitudes and dispositions—regarding math, science, and technology. Furthermore, research on the role of social processes in early learning in mathematics, science, and technology makes moot the presumed need to choose between responding to children's social needs versus their intellectual or academic needs. For example, studies on children's early development of number concepts illustrates the interplay among social, intellectual, and developmental processes (Saxe et al. 1987). That children see the personal relevance of what they are learning, and receive appropriate social support, is critical to their formation of mathematical concepts (Ball and Wilson 1996) and to their development of interest in science (Jeffe 1995) and computers (Char and Forman 1994). Indeed, when children work together on a computer, their social exchanges promote not only their learning of technological skills (Clements 1994), but they can also facilitate their use of the computer to acquire advanced understandings of literacy, mathematics, and science (Wright and Shade 1994).

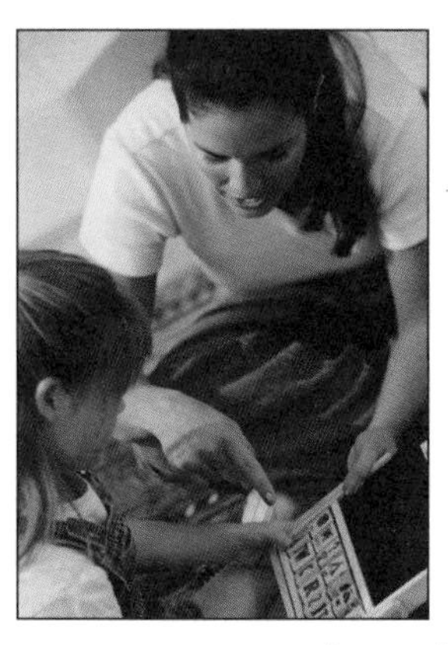

The benefits of social negotiations among students as they take place within collaborative learning have been demonstrated in research on the teaching and learning of science (Fosnot 1996) and mathematics (Saxe and Gearhart 1988). Such studies support theoretical understandings of learning as both an individual and a social process (Shapiro 1994). They also support new interpretations of the domains of mathematics and science themselves, where knowledge is negotiated through social exchanges within particular socio-cultural contexts (Forman 1993).

Research informed by social constructivism also supports the role of peers and teachers in facilitating instruction in mathematics, science, and technology. It also suggests that appropriate educational opportunities in these domains can enhance other aspects of children's development. For example, we now know that children with emotional or behavioral disabilities can learn about cause and effect in their joint science activities with more capable children. Children with cognitive impairments benefit from experiences that require active thinking and reasoning about problems (including scientific and mathematics problems) that matter to them. Students with physical or sensory impairments are highly motivated to use all of their available senses in order to better observe natural phenomena (Mastropieri and Scruggs 1995). This body of research supports the notion that children of all abilities take clues from the physical and social environment regarding what is important to learn and how it might be learned (Mallory and New 1994b).

These advances in our understanding of how children learn have significant implications for the role of the early childhood educator in the early childhood curriculum. Revised interpretations of developmentally appropriate practice (Bredekamp and Copple 1997) now make explicit reference to the critical importance of teacher observations about (1) what children know and are ready to learn and (2) the nature of various forms of teacher assistance that will facilitate the child's exploration with new materials, concepts, and conflicts. The theoretical concept of guided participation has blurred the distinction between teacher-directed and child-sensitive pedagogy. The value of teacher promotion of conceptual understanding is no longer seen as dichotomous to the role of play in children's learning and development.

## Reconceptualization of the Early Childhood Curriculum

Recent descriptors of the early childhood curriculum include integrated (with respect to developmental goals), emergent (with respect to the source of content or theme), and negotiated (as opposed to either teacher- or child-initiated). Each of these interpretations of developmentally appropriate curriculum includes the belief that children are considerably more likely to achieve goals that adults set for them when the content of new knowledge is personally meaningful, is contextually relevant, and builds upon, rather than replaces, existing competencies (New 1998a). These expectations for the curriculum place a heavy emphasis on the role of the teachers, who have the responsibility of ensuring that children have opportunities to learn from one another, that they have ample motivation to revisit their understandings, and that they are encouraged to reflect critically on their own and each other's ideas. Such a curriculum also requires that teachers, too, see themselves as students of children's learning and development. Teachers in Reggio Emilia, Italy, have done much to help clarify these points (Edwards et al. 1993; New 1998b).

Expanded conceptions of developmentally appropriate practice have responded to the need to acknowledge the diversity of practices that may be appropriate for diverse populations of young children (Mallory and New 1994a). Current thinking also emphasizes the importance of connecting curriculum content with the larger context in which children live. Experiences with mathematical concepts, scientific problem solving, and computer technology can relate to other aspects of children's lives. These experiences can also create occasions for children to think critically, make predictions, and solve problems.

The challenge in promoting competence in the skills and knowledge deemed critical by the larger culture is to consider the usefulness of such knowledge from the perspective of children (and their families) who are culturally or linguistically diverse. Children whose family lives are outside the mainstream ought to be encouraged to explore and express their own specialized knowledge (Phillips 1994). They must also be viewed as entitled to have access to opportunities and resources otherwise unavailable (Delpit 1995). For children attempting to bridge two worlds, the role of the teacher is to embrace both realities and to model the acceptance of competence in its diverse forms and origins.

Discussions of equity in mathematics, science, and technology are typically limited to consideration of the fairness of access and opportunities to participate in activities related to those domains. However, the recent reconceptualization of the early childhood curriculum also utilizes mathematics, science, and technology to address attitudes and practices associated with issues of equity. Science, for example, provides a wonderful opportunity to utilize cooperation and problem-solving skills as small groups of children test their capacities to generate and test hypotheses. As children engage in scientific processes of observation, hypothesis generating, and hypothesis testing, they can be challenged to confront their own understandings with those of their peers.

Children struggling to utilize mathematics concepts to make classroom decisions can also be encouraged to consider the extent to which numerical advantage translates into fair play. For example, what does it mean to divide, to share, to be fair. When is the voting process not democratic? Under what conditions does a 16-to-7 outcome silence a minority voice that should be heard? Such critical analysis serves to promote children's comprehension of the conceptual bases of mathematical computation as well as their efforts to disentangle numerical worth from social meanings (Ball and Wilson 1996). In Reggio Emilia, Italy, an athletic project on the long jump ultimately inspired children to debate the nature of gender competencies, the mathematical interpretation of a handicap, and a friendly means of taking into account different competencies when comparing distances achieved by boys and girls of different ages and abilities.

Far too often, teachers presume that children have neither the interest nor the ability to respond to socially complex issues. In fact, some of the children's most serious engagement takes place when they pursue moral dilemmas behind the observation, "Our school's not fair!" (Pelo 1997). Recent interpretations of the social foundations of cognition emphasize the critical role of the classroom in promoting vigorous and respectful engagement around topics of social and intellectual significance (Tharp and Gallimore 1988). Such experiences can contribute not only to conceptual changes in scientific and mathematical thinking, but they can also increase children's appreciation of the relevance of mathematics and science to their daily lives. They can begin to view these topics as a means of improving their own thinking and their relationships to each other.

## Conclusion

This reconceptualization of the early childhood curriculum—and the teaching of mathematics, science, and technology—is based not only on new understandings of how children learn, but on what they need to learn for life in a pluralistic, democratic society. And just as it is increasingly vital that children acquire conceptual understandings in mathematics, science, and technology, so too is it essential that children begin to comprehend the role that such knowledge plays in a contemporary democratic society. While many might claim that such learning goals are far from the reach of three-, four- and five-year-old children, the reconceptualized early childhood curriculum shares the belief that, just because teachers ought to begin where children are, "Beginning there has never implied staying there" (Wright 1965, page 34).

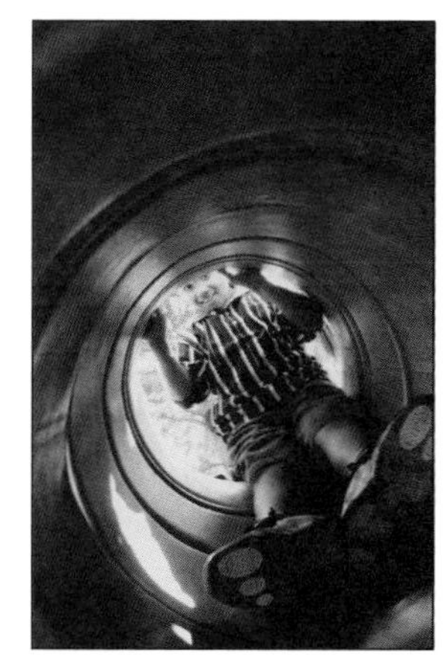

Nineteen years ago it was suggested that educators already knew enough to successfully teach all children and that it was primarily a question of "how we feel about the fact that we haven't so far" (Edmonds 1979, page 22). Perhaps this confidence in the knowledge base of the profession was prematurely optimistic, given the changes in our understandings since that time. Today, however, it does seem that we know a great deal about what we ought to be doing better.

Early childhood educators have the opportunity to make an immediate difference in at least a portion of the life (several hours a day) of many young children. Successful early childhood programs have also demonstrated their potential to make a difference in the continuing lives of the children and their families. "Programs that work" for minority and impoverished children in the United States act upon the theoretical premise that it is necessary to connect with children's lives; on the political premise that it is critical to advocate for their well-being; and on the ethical premise that it is essential to contribute to parents' abilities to support the learning and development of their children (Barnett and Boocock 1998). Families of young children must be involved in deciding upon and incorporating educational goals in mathematics, science, and technology into the early childhood curriculum. Such a change in the standard home-school relationship will require more than an increase in teachers' professional development and parent education activities. It will require a change in attitude regarding the collective responsibility for the education of young children in an inequitable society.

The bigger question of equity in educational resources and opportunities remains a dilemma. Simply acknowledging, as a society, that the problem continues to exist may be one of the greatest challenges. At minimum, educators, community members, policy makers, and other taxpayers must somehow face up to the fact that Goodlad's 1984 comment that schools "mirror inequities in the surrounding society and many people want to be sure that they continue to do so" remains no less true today (Goodlad 1984). Recent analyses of school reform efforts reveal the difficulties in eliminating tracking and other forms of segregation when elite groups of parents insist on maintaining such special distinctions as gifted programs for their children. This ideology of "diversity at a distance" (Wells and Serna 1996) threatens any meaningful effort to close the education gap in American society.

This paper began by acknowledging the period of early childhood as ripe for development and vulnerable to neglect. The societal context of inequities in the classroom suggests that the goal of achieving more equitable and effective means for teaching children particular subject matter (i.e., mathematics, science, and technology) is directly linked to our society's willingness to express a more collective commitment to all young children. As we approach the end of the 20th century, we are witness to an explosion of knowledge about children's real and potential competencies—and the consequences of their neglect. Perhaps the biggest challenge of the next century will be to actualize our potentials as adults by better advocating on children's behalf.

## References and Bibliography

American Association for the Advancement of Science. (1994). *Benchmarks in science literacy.* Washington, DC: Author.

Ball, D.L., and Wilson, S.M. (1996). Integrity in teaching: Recognizing the fusion of the moral and intellectual. *American Educational Research Journal,* 33(1): 155–192.

Barnett, W.S., and Boocock, S.S., eds. (1998). *Early care and education for children in poverty: Promises, programs, and long-term results.* Ithaca, NY: SUNY Press.

Beatty, B. (1995). *Preschool education in America: The culture of young children from the colonial era to the present.* New Haven: Yale University Press.

Berk, L.E., and Winsler, A. (1995). *Scaffolding children's learning: Vygotsky and early childhood education.* Washington, DC: National Association for the Education of Young Children (NAEYC).

Bredekamp, S. (1987). *Developmentally appropriate practice in early childhood programs serving children from birth through age 8.* Washington, DC: NAEYC.

Bredekamp, S., and Copple, C., eds. (1997). *Developmentally appropriate practice in early childhood programs.* (Rev. ed.). Washington, DC: NAEYC.

Char and Forman, G. (1994). Interactive technology and the young child: A look to the future. In *Young children: Active learners in a technological age,* eds. J.L. Wright and D.D. Shade. Washington, DC: NAEYC.

Clements, D. (1994). The uniqueness of the computer as a learning tool: Insights from research and practice. In *Young children: Active learners in a technological age,* eds. J.L. Wright and D.D. Shade. Washington, DC: NAEYC.

Delpit, L. (1995). *Other people's children: Cultural conflict in the classroom.* New York: The New York Press.

Derman-Sparks, L., and the A.B.C. Task Force. (1989). *Anti-bias curriculum: Tools for empowering young children.* Washington, DC: NAEYC.

Dewey, J. (1911). Is coeducation injurious to girls? *Ladies Home Journal,* June 11, 1911: 60–61.

Edmonds, R. (1979). Effective schools for the urban poor. *Educational Leadership*, 37(1): 15–24.

Edwards, C., Gandini, L., and Forman, G., eds. (1993). *The hundred languages of children: The Reggio Emilia approach to early childhood education*. (Rev. 2nd ed.). Norwood, NJ: Ablex.

Fennema, E., Franke, M.L., Carpenter, T.P., and Carey, D.A. (1993). Using children's mathematical knowledge in instruction. *American Educational Research Journal*, 30(3): 555–583.

Forman, G. (1993). The city in the snow: Applying the multisymbolic approach in Massachusetts. In *The hundred languages of children: The Reggio Emilia approach to early childhood education*, eds. C. Edwards, L. Gandini, and G. Forman. Norwood, NJ: Ablex.

Fosnot, C., ed. (1996). *Constructivism: Theory, perspectives, and practice*. New York: Teachers College Press.

Ginsburg, H.P., and Asmussen, K.A. (1988). Hot mathematics. In *Children's mathematics, new directions for child development*, No. 41, eds. G.B. Saxe and M. Gearhart. San Francisco: Jossey-Bass.

Goldhaber, J. (1994). If we call it science, then can we let the children play? *Childhood Education*, 24–27.

Goodlad, J.L. (1984). *A place called school*. New York: McGraw-Hill.

Goodman, M.E. (1952). *Race awareness in young children*. New York: Crowell-Collier.

Harry, B. (1992). *Cultural diversity, families, and the special education system: Communication and empowerment*. New York: Teachers College Press.

Jeffe, D.B. (1995). About girls' "difficulties" in science: A social, not a personal, matter. *Teachers College Record*, 97(2): 206–226.

Kadden, M. (1990). Issues on computers and early childhood education. In *Continuing issues in early childhood education*, ed. C. Seefeldt, 261–275. Columbus, OH: Merrill.

Kagan, S.L., and Cohen, N., eds. (1996). *Reinventing early care and education: A vision for a quality system*. San Francisco: Jossey-Bass.

Kendell, F.E. (1996). *Diversity in the classroom: New approaches to the education of young children*. New York: Teachers College Press.

Lewis, C. (1995). *Educating hearts and minds: Reflections on Japanese preschools*. Cambridge, UK: Cambridge University Press.

Mallory, B.L. (1998). Educating young children with developmental differences: Principles of inclusive practice. In *Continuing issues in early childhood education*, eds. C. Seefeldt and A. Galper, 213–237. Columbus, OH: Merrill.

Mallory, B.L., and New, R.S. (1994a). *Diversity and developmentally appropriate practices: Challenges to early childhood education.* New York: Teachers College Press.

Mallory, B., and New, R. (1994b). Social constructivist theory and principles of inclusion: Challenges for early childhood special education. *Journal of Special Education*, 28(3): 322–337.

Mastropieri, M., and Scruggs, T. (1995). Teaching science to students with disabilities in general education settings. *Teaching Exceptional Children*, 27(4): 10–13.

Meece, J.L. (1987). The influence of school experiences on the development of gender schemata. In Children's gender schemata, *New directions for child development*, eds. G.B. Saxe and M. Gearhart. San Francisco: Jossey-Bass.

New, R. (1998a). Diversity and early childhood education: Making room for everyone. In *Continuing issues in early childhood education*, eds. C. Seefeldt and A. Galper, 238–268. Columbus, OH: Merrill.

New, R. (1998b). Theory and praxis in Reggio Emilia: They know what they are doing and why. In *The hundred languages of children: The Reggio Emilia approach to early childhood education*, (Rev. 2nd ed.), eds. C. Edwards, L. Gandini, and G. Forman. Norwood, NJ: Ablex.

New, R., and Mallory, B. (1996). The paradox of diversity in early care and education. In *Putting children first: Visions for a brighter future for young children and their families*, ed. E. Erwin. Baltimore, MD: Brookes.

Oakes, J. (1985). *Keeping track: How schools structure inequality.* New Haven, CT: Yale University Press.

Pelo, A. (1997). "Our school's not fair!" A story about emergent curriculum. *Young Children*, 52(7): 57–61.

Phillips, C. (1994). The movement of African-American children through socio-cultural contexts: A case of conflict resolution. In *Diversity and developmentally appropriate practices: Challenges for early childhood education*, eds. B.L. Mallory and R.S. New. New York: Teachers College Press.

Saxe, G.B., and Gearhart, M. (1988). *Children's mathematics: New directions for child development.* San Francisco: Jossey-Bass.

Saxe, G.B., Guberman, S.R., and Gearhart, M. (1987). Social processes in early number development. *Monographs of the Society for Research in Child Development*, 52(2), Serial No. 216.

Shapiro, B. (1994). *What children bring to light: A constructivist perspective on children's learning in science.* New York: Teachers College Press.

Tharp, R., and Gallimore, R. (1988). *Rousing minds to life.* Cambridge, England: Cambridge University Press.

Wells, A.S., and Serna, I. (1996). The politics of culture: Understanding local political resistance to detracking in racially mixed schools. In *Working together toward reform.* Cambridge, MA: Harvard College. Reprinted from *Harvard Educational Review,* 66(1): 93–118.

Woodhead, M. (1996). *In search of the rainbow: Pathways to quality in large-scale programmes for young disadvantaged children.* The Hague: Bernard Van Leer Foundation.

Wright, B.A. (1965). *Educating for diversity.* New York: John Day Co.

Wright, J.L., and Shade, D.D., eds. (1994). *Young children: Active learners in a technological age.* Washington, DC: NAEYC.

# Selected Resources

# Selected Resources

## Activities and Resources for Parents

Allison, Linda, and Martha Weston. 1994. Pint-Size Science: Finding-out Fun for You and Your Young Child. Little Brown. Waltham, MA. ISBN 0–31603–467–3.

Hands-on science activities. A Brown Paper Preschool Book.

Borden, Marian Edelman. 1997. Smart Start: The Parents' Complete Guide to Preschool Education. Facts on File. New York, NY. ISBN 0–81603–604–7.

Helps parents identify high-quality programs that provide preschoolers opportunities for successful learning experiences. Combines input from psychologists, preschool educators, and pediatricians with the experiences, stories, and insights of parents.

Carlson, Laurie. 1995. Green Thumbs. Chicago Review Press. Chicago, IL. ISBN 1–55652–238–X.

Hands-on nature activities that may be done indoors or outdoors. Activities include keeping a garden diary and making herbal soap and flower petal candy.

Cohen, Richard. 1997. Snail Trails and Tadpole Tails: Nature Education for Young Children. Redleaf Press. St. Paul, MN. ISBN 0–93414–078–2.

Hands-on science activities to teach nature to children.

ERIC Clearinghouse. 1997. Parent Brochures. ERIC Clearinghouse on Elementary and Early Childhood Education. Champaign, IL. http://ericeece.org

Site offers more than 30 links to online brochures for parents on topics ranging from "What can I teach my young child about the environment?" to "What does school reform mean to my neighborhood school?" Provides links to all ERIC sites and the ERIC publications catalog.

Goin, Kenn, Eleanor Ripp, and Kathleen Nastasi Solomon. 1997. Bugs to Bunnies: Hands-on Animal Science Activities for Young Children. Gryphon House. Beltsville, MD. ISBN 0–94312–903–6.

Interdisciplinary hands-on activities about animals. Learn about camouflage, spider webs, and baleen whales, among others.

Harlan, Jean Durgin, and Carolyn Good Quattrocchi . 1994. Science As It Happens. Henry Holt & Company. New York. ISBN 0–8050–3061–1.

Family activities with children ages 4 to 8 that teach basic principles about the way the world works.

Kanter, Patsy F. 1992. Helping Your Child Learn Math. U.S. Department of Education, Office of Educational Research Improvement. Washington, DC.

Activities for children aged 5 to 13.

Katz, Lilian. 1995. How Can Parents Identify a High Quality Preschool Program? ERIC Clearinghouse on Elementary and Early Childhood Education. Champaign, IL. http://ericeece.org

Offers parents tips on choosing a preschool program for their children. Includes descriptions of different needs of children and different types of programs.

Katz, Lilian G. 1987. What Should Young Children Be Learning? http://ericeece.org

Addresses such issues as appropriate teaching curriculum approaches and where parents can find out more about kindergarten practices.

Kohl, MaryAnn F., and Cindy Gainer. 1996. MathArts. Gryphon House. Beltsville, MD. ISBN 0–87659–177–2.

Hands-on math activities using creative arts to teach early math concepts such as one-to-one correspondence, sorting, grouping, classifying, opposites, number values, and number recognition. Selected activities available online at http://www.ghbooks.com.

Miami University and National Science Teachers Association. 1998. Dragonfly. NSTA. Arlington, VA. ISSN 1809–9006.

Online and print subscription magazine offering varieties of articles, photographs, and hands-on activities that introduce young children to scientific facts. Themes include animal communication, flight, weather, and trees. Links allow readers to e-mail researchers, submit articles, find out about topics in upcoming issues, move to related sites, and more. Available at http://www.nsta.org/pubs/dragonfly.

Milford, Susan. 1995. The Kid's Nature Book. Gryphon House. Beltsville, MD. ISBN 0–91358–94–X.

Hands-on science activities about nature. Activities range from preparing a butterfly garden to making wind chimes to planting a bottle garden to hatching amphibian eggs.

Montessori Foundation. 1998. Tomorrow's Child. Montessori Foundation. Alexandria, VA. ISSN 1071–6246.

Magazine written specifically for parents with children enrolled in Montessori schools.

Moomaw, Sally, and Brenda Heironymous. 1995. More than Counting: Exploring Math Activities for Preschool and Kindergarten. Gryphon House. Beltsville, MD. ISBN 0–93414–082–0.

Interactive math activities using manipulatives, grids, games, gross-motor skills, and graphing.

Moomaw, Sally, and Brenda Heironymous. 1997. More than Magnets: Exploring the Wonders of Science in Preschool and Kindergarten. Gryphon House. Beltsville, MD. ISBN 1–88483–433–7.

Hands-on physical science activities.

Redleaf, Rhoda. 1996. Open the Door, Let's Explore More! Gryphon House. Beltsville, MD. ISBN 1–88483–413–2.

Includes songs, fingerplays, and resource lists to encourage exploration beyond the classroom.

Reed, Ruth Barnes. 1996. Computers in the Home. Tomorrow's Child. The Montessori Foundation. Alexandria, VA.

An article supported by The Montessori Foundation that offers ways to make computer use at home an exciting and innovative way for children to learn. Available online at http://www.montessori.org.

Rockwell, Robert E., Elizabeth A. Sherwood, and Robert A. Williams. 1992. Everybody Has a Body: Science from Head to Toe. Redleaf Press. St. Paul, MN. ISBN 0–87659–158–6.

Hands-on science activities about the body. Promotes skills such as observation, inference, and prediction. Selected hands-on activities available online at http://www.ghbooks.com.

Ross, Michael. 1996. Sandbox Scientist. Gryphon House. Beltsville, MD. ISBN 1–55652–248–7.

Open-ended, hands-on science activities including ice and bubbles, compost and seeds, magnets and gears, potions and plant prints.

Seldin, Tim. 1996. The Inner Life of the Child. Tomorrow's Child. The Montessori Foundation. Alexandria, VA.

An article by the president of the Montessori Foundation that describes his impressions of the inner life of the child. Available online at http://www.montessori.org.

Sherwood, Elizabeth A., Robert E. Rockwell, and Robert A. Williams. 1990. More Mudpies to Magnets: Science for Young Children. Gryphon House. Beltsville, MD. ISBN 0–87659–150–0.

Hands-on science activities and ideas. Builds on skills such as classification, measurement, and prediction. Selected activities available online at http://www.ghbooks.com.

Swanson, Beverly B. 1997. What Is a Quality Preschool Program? ERIC Clearinghouse on Elementary and Early Childhood Education. Champaign, IL. http://ericeece.org

Offers parents tips on choosing a preschool program for their children.

Townsend-Butterworth, Diana. 1995. Preschool and Your Child: What You Should Know. Walker & Co. New York, NY. ISBN 0–80277–472–5.

A parent's guide to 10 key ingredients in choosing the best preschool program for any child. Lists questions for parents to ask schools.

U.S. Department of Education. 1991. Helping Your Child Learn Science. U.S. Department of Education. Washington, DC.

Includes articles and hands-on science activities appropriate for use in the home and in community programs.

Warner, Penny. 1996. Splish, Splash. Chicago Review Press. Chicago, IL. ISBN 1–55652–262–2.

Hands-on science activities that teach about the wonders of water. Activities include Musical Mud Puddles and Water Whirlers.

## Assessment in Mathematics, Science, and Technology

Bagnato, Stephen J., John T. Neisworth, and Susan M. Munson. 1997. Linking Assessment and Early Intervention: An Authentic Curriculum-Based Approach. Paul H. Brooks Publishing Co. Baltimore, MD. ISBN 1–55766–263–0.

The authors apply a six-standard index to more than 50 different curriculum-embedded and curriculum-compatible assessment and intervention systems. They explain how early childhood professionals can perform reviews and customize systems to specific programs and populations.

Bergan, John R., and Jason K. Feld. 1993. Developmental Assessment: New Directions. Young Children, Vol. 48, No. 5. NAEYC. Washington, DC. ISSN 0044–0728.

Discusses the Management and Planning System, a developmental assessment system initiated in Head Start. The system consists of instruments that assess preschoolers' and kindergarteners' development of math, science, literacy, social skills, and motor development.

Bridgeman, Brent, Edward Chittenden, and Frederick Cline. 1994. Characteristics of a Portfolio Scale for Rating Early Literacy. Educational Testing Service. Princeton, NJ.

Brochure discusses assessment in the early years.

Meisels, Samuel J. 1996. Using Work Sampling in Authentic Assessment. Educational Leadership, Vol. 54, No. 4. Association for Supervision and Curriculum Development. Alexandria, VA. ISSN 0013–1784.

Early childhood and elementary school teachers are using this authentic performance assessment to document children's learning experiences, meet standards, and connect assessment to instruction.

## Cognitive and Developmental Science

Beaty, Janice J. 1997. Observing Development of the Young Child, 4th edition. Merrill Publishing Co. New York, NY. ISBN 0–13801–986–X.

Introduction to development in children ages 2 to 6, including observing and assessing their behavior. Includes guidelines for appropriate assessment of young children.

Bredekamp, Sue, and C. Copple (Eds.). 1997. Developmentally Appropriate Practice in Early Childhood Programs. NAEYC. Washington, DC. ISBN 0–93598–979–X.

Carpenter, T.P., et al. 1993. Models of Problem Solving: A Study of Kindergarten Children's Problem-Solving Processes. Journal for Research in Mathematics Education, Vol. 24, No. 5. NCTM. Reston, VA. ISSN 0021–8251.

Copley, Juanita. 1998. Mathematics in the Early Years, Ages 0–5. NCTM. Reston, VA.

Crosser, Sandra. 1996. The Butterfly Garden: Developmentally Appropriate Practice Defined. Early Childhood News, Vol. 9, No. 2. Peter Li Education Group. Dayton, OH. ISSN 1080–3564.

Presents an overview of what is meant by a developmentally appropriate preschool classroom. Full text available online at http://www.earlychildhoodnews.com

Day, Barbara. 1994. Early Childhood Education: Developmental/Experiential Teaching and Learning, 4th edition. Macmillan. London. ISBN 0–02327–923–0.

Examines developmentally appropriate learning centers and how they are best managed and organized to meet the needs of young children, including those with special needs. Variety of activities provided.

Devonshire, Hilary. 1991. Water Science Through Art. Grolier Publishing. Danbury, CT. ASIN 053114125X.

This excellent book first defines, then demonstrates certain properties of water as art media: Water flows, is wet, evaporates, can be absorbed, and can form crystals.

Dixon, Dorothy and Carole MacClennan. 1995. See How They Grow: The Early Childhood Years. Twenty-Third Publications. ISBN 0–89622–567–4.

Dunn, Loraine, and Susan Kontos. 1997. What Have We Learned about Developmentally Appropriate Practice? Young Children, Vol. 52, No. 5. NAEYC. Washington, DC. ISSN 0044–0728.

Assesses what we have learned from a decade of research on developmentally appropriate practice.

Elkind, David. 1989. The Hurried Child: Growing Up Too Fast Too Soon, Revised edition. Addison-Wesley Publishing Co. Reading, MA. ISBN 0–20107–398–8.

Examines the stresses on children forced to grow up too fast, children who mimic adult sophistication while secretly yearning for innocence.

Fleer, Marilyn. 1996. Play Through the Profiles: Profiles Through Play. Australian Early Childhood Association. Watson, Australian Capital Territory, Australia. ISBN 1–87589–020–3.

Validates play as a fundamental component of learning and an avenue through which learning outcomes can be identified and confirmed. Supports science and math learning through play.

Gardner, Howard. 1991. The Unschooled Mind: How Children Think and How Schools Should Teach. Basic Books. ISBN 0–46508–895–3.

Reviews child development theories and provides a description of schooling based on apprenticeship and museum learning models. Summarizes recent research on children's conceptions in science, math, and other areas.

Hirschfeld, Lawrence A., and Susan A. Gelman (Eds.). 1994. Mapping the Mind: Domain Specificity in Cognition and Culture. Cambridge University Press. Portchester, NY. ISBN 0–52141–966–2.

Many researchers have concluded that much of human thought is "domain specific." Consequently, the mind is better viewed as a collection of cognitive abilities specialized to handle specific tasks. This introduction explores how these cognitive abilities are organized.

Holmes, Madelyn (Ed.). 1997. Early Learning. Basic Education, Vol. 41, No. 9. Council for Basic Education. Washington, DC. ISSN 1964–984.

New information about how the brain develops may lead educators to rethink how they maximize every child's opportunity to learn. Brief articles examine how the brain works, early education, and exemplary preschool programs.

Leushina, A.M. 1991. The Development of Elementary Mathematical Concepts in Preschool Children. NCTM. Reston, VA. ISBN 0–87353–299–6.

Discusses six stages in the development of counting and explains children's notions of size, shape, mass, measurement, spatial orientation, and time. Originally published in 1974, it is based on research conducted in the former Soviet Union.

Resnick, Lauren B. 1987. Education and Learning To Think. National Academy Press. Washington, DC. ISBN 0–60802–330–2.

Report addresses what American schools can do to more effectively teach higher order thinking skills. Full text available online at http://www.ul.cs.cmu.edu.

Seefeldt, Carol, and Alice Galper. 1998. Continuing Issues in Early Childhood Education. Prentice-Hall, Inc. Upper Saddle River, NJ. ISBN 0–13519–364–8.

Spodek, Bernard (Ed.). 1993. Handbook of Research on the Education of Young Children. Macmillan. ISBN 0–02897–405–0.

Sternberg, R.J., and T. Ben-Zeev (Eds.). 1996. The Nature of Mathematical Thinking. Lawrence Erlbaum Associates. Mahwah, NJ. ISBN 0–80581–798–0.

Williams, R., R. Rockwell, and E. Sherwood 1987. Mudpies to Magnets. Gryphon House. Beltsville, MD.

Includes directions for making bleach-bottle "space helmets" and a balance toy.

Wood, Chip. 1997. Yardsticks: Children in the Classroom Ages 4–14: A Resource for Parents and Teachers, 2nd edition. Northeast Foundation for Children. Greenfield, MA. ISBN 0–06186–364–1.

Provides teachers and parents with a reference on important childhood development issues, explaining what children should be learning and doing in the classroom at each developmental stage.

## Early Childhood Education Curriculum Materials and Instruction

Adams, Polly K., and Jaynie Nesmith. 1996. Blockbusters: Ideas for the Block Center. Early Childhood Education Journal, Vol. 24, No. 2. Human Sciences Press. New York, NY. ISSN 1082–3301.

Goals of block building in early childhood classrooms focus on physical, social, cognitive, and emotional development. Offers illustrations of task cards to use with blocks in math, science, language arts, and social studies.

Atkinson, Sue, and Marilyn Fleer. 1996. Science with Reason: A Developmental Approach. Heinemann. Portsmouth, NH. ISBN 0–43508–381–3. http://www.heinemann.com.

Based on the premise that learning is best accomplished when children engage in purposeful activity and real problems, 16 contributors describe what this kind of science teaching can look like for a range of science topics for primary school and elementary school children.

Baker, Ann. 1991. Counting on a Small Planet: Activities for Environmental Mathematics. Heinemann. Portsmouth, NH. ISBN 0–43508–327–9

The "whole math" approach is explained, and ideas are suggested for getting children to use math naturally when exploring their environment.

Beaty, Janice J. 1996. Preschool: Appropriate Practices, 2nd edition. Harcourt Brace. San Diego, CA. ISBN 0–15502–633–X.

This teacher's manual focuses on developing self-directed "learning environments" for preschool children and on facilitating the children's development through observation and support. Describes how to set up "centers" in classrooms, including those that focus on computers, math, and science.

Berry, Carla F. 1992. Planning a Theme-Based Curriculum: Goals, Themes, Activities, and Planning Guides for 4's and 5's. Addison-Wesley Publishing Co. Reading, MA. ISBN 0–67346–409–1.

Broekel, Ray. 1988. Experiments With Water. Children's Press. Danbury, CT. ISBN 0–51601–215–0.

Experiments with temperature changes, capillary action, surface tension, and buoyancy are provided for children ready for additional challenges. (Photographs show children handling equipment that should be used under adult supervision.)

Bugs Don't Bug Us. Bo Peep Productions. 1991. http://www.bopeepproductions.com

This video features insects in action in their natural settings, even changing into mature forms, plus creative movements and a good song that will build a child's courage about encountering these creatures.

Burns, Marilyn. 1998. Thinking Math: Questions To Ask and Games To Play To Help Children Think Mathematically. Early Childhood Today, Vol. 12, No. 4. Scholastic, Inc. Jefferson City, MO. ISSN 1070–1214.

Caduto, Michael. 1990. Pond and Brook: A Guide to Nature Study in Freshwater Environments. University Press of New England. ISBN 0–87451–509–2.

Many creative ideas for involving children with nature.

Charlesworth, Rosalind. 1996. Experiences in Math for Young Children, 3rd edition. Delmar Publishers. Albany, NY. ISBN 0–82737–227–2.

Developmentally appropriate integrated curriculum is stressed as well as language and literature.

Claycomb, Patty. 1991. Love the Earth. Partner Press. ASIN 0–93321–247–X.

Many creative ideas for involving children with nature.

Eisenhower National Clearinghouse. 1998. Using Children's Literature in Math and Science. ENC Focus. Eisenhower National Clearinghouse. Columbus, OH.

Helps teachers integrate reading and math and science, particularly in the elementary grades. Lists trade books to use in math and science as well as other resources.

ERIC Clearinghouse/EECE. A to Z: The Early Childhood Educator's Guide to the Internet. ERIC Clearinghouse on Elementary and Early Childhood Education. Champaign, IL. http://ericeece.org.

Offers an introduction to the Internet, describes common discussion-list commands, lists web sites, and provides instruction on finding and using ERIC on the Internet.

GEMS, Lawrence Hall of Science. 1993. Once Upon a GEMS Guide: Connecting Young People's Literature to Great Explorations in Math & Science. GEMS, Lawrence Hall of Science. Berkeley, CA. ISBN 0–91251–178–8.

Annotates hundreds of books that can be read in connection with GEMS activities, math strands, and science themes for PreK–12.

George, Yolanda S., et al. (Eds.). 1995. In Touch With Preschool Science. American Association for the Advancement of Science. Washington, DC. ISBN 0–87158–551–5.

Includes information on starting a preschool science program and a wide selection of activities, written in both English and Spanish, to do with young children and their families.

Green, Moira D. 1996. 474 Science Activities for Young Children. Delmar Publishers. Albany, NY. ISBN 0–8273–663–9.

Child-initiated science projects incorporate "whole language" learning with a multicultural and anti-bias foundation.

Greenberg, Polly. 1993. Ideas That Work with Young Children: How and Why to Teach All Aspects of Pre-K, K Math Naturally. Young Children, Vol. 48, No. 4. NAEYC. Washington, DC. ISSN 0044–0728.

Teachers who are not mathematically inclined can learn enough math and enough about young children to become more satisfactory teachers of mathematics.

Hampton, Carol, and David Kramer. 1994. Classroom Creature Culture: Algae to Anoles. National Science Teachers Association. Arlington, VA. ISBN 0–87355–120–6.

This anthology of articles from *Science and Children* magazine focuses on the care of plants and animals brought into school from the wild. It is an important resource for teachers.

Horenstein, Sidney. 1993. Rocks Tell Stories. Houghton Mifflin. Boston. ISBN 0–39566–818–2.

The level of concise information in this book for older children makes good background prepartion for teachers of young children.

Levenson, Elaine. 1994. Teaching Children About Physical Science. McGraw-Hill. New York. ISBN 0–07037–619–0.

Lind, Karen K. 1998. Exploring Science In Early Childhood: A Developmental Approach, 2nd edition. Delmar Publishers. Albany, NY. ISBN 0–82737–309–0.

Contains research and experiments to help teachers develop appropriate science curriculum for preschool and primary-age children.

Macaulay, David. 1988. The New Way Things Work: A Visual Guide to the World of Machines. Houghton Mifflin. Boston. ISBN 0–39593–847–3.

This book gives the teacher the same revelations of "so that's how it works" that children gain from their study of simple machines.

McCarty, Diane, et al. 1996. Mini-Portfolio on Math and Science. Teaching PreK–8, Vol. 26, No. 4. Early Years, Inc. Norwalk, CT. ISSN 0891–4508.

Presents six articles dealing with math and science education.

McIntyre, Margaret. 1984. Early Childhood and Science. NSTA. Arlington, VA. ISBN 0–87366–029–3.

This collection from *Science and Children* magazine offers suggestions for integrating the science skills of observation, identification, and exploration with traditional activities in art, music, and literature.

McVey, Vicki. 1991. The Sierra Club Book of Weatherwisdom. Sierra Club Books. San Francisco. ASIN 0-31656-341-2.

General background information on weather, plus suggestions for activities that could be adapted for early childhood classrooms.

Montessori Foundation. 1996. A Montessori Curriculum Scope and Sequence, Age 3–12. Montessori Foundation. Alexandria, VA.

Geography, mathematics, geometry, and life sciences are incorporated in this paper, which explains the curriculum for Montessori schools.

Moving Machines. Bo Peep Productions. http://www.bopeepproductions.com.

In this video, parallels are shown between heavy construction machines at work and young children playing with toys that look like these machines.

National Science Resources Center. 1996. Resources for Teaching Elementary School Science. National Academy Press. Washington, DC. ISBN 0–30905–293–9.

Annotated guide to hands-on, inquiry-centered curriculum materials and sources of help for teaching science from K–6.

National Science Resources Center. 1997. Science for All Children: A Guide to Improving Elementary Science Education in Your School District. National Academy Press. Washington, DC. ISBN 0–30905–297–1.

Oppenheim, Carol. 1993. Science is Fun! Delmar Publishers. Albany, NY. ISBN 0–82737–337–6.

Includes nature and science facts and activities appropriate for children ages 2 through 8. Stresses the importance of enjoying and appreciating our world.

Parker, Steve. 1992. The Random House Book of How Nature Works. Random House. New York. ASIN 0–67983–700–0.

Excellent background information on how animals meet their needs for food and air and how they grow, move, and protect themselves. Detailed illustrations. Paperback.

Pellant, Chris. 1992. Rocks and Minerals. Dorling Kindersley. London. ISBN 1–56458–033–4.

The excellent color photographs in this *Eyewitness Handbook* make it possible to identify any rock brought in by children, but the descriptions are highly technical.

Perry, Gail, and Mary Rivkin. 1992. Teachers and Science. Young Children, Vol. 47, No. 4. NAEYC. Washington, DC. ISSN 0044–0728.

Offers readers two perspectives—that of children and that of teachers–on getting started on good science. Suggestions are given to help those working with young children become better teachers of science.

Perry, Phyllis J. 1996. Rainy, Windy, Snowy, Sunny Days: Linking Fiction to Non-Fiction. Teacher Ideas Press. Englewood, CO. ISBN 1–56308–392–2.

Rivkin, Mary. 1992. Science Is a Way of Life. Young Children, Vol. 47, No. 4. NAEYC. Washington, DC. ISSN 0044–0728.

Article discusses how science is a way of doing things and solving problems. It is a style that leads a person to wonder, to seek, to discover, and then to wonder anew.

Russell, Helen Ross. 1991. Ten-minute Field Trips: Using the School Grounds for Environmental Studies (2nd ed.). National Science Teacher's Association. Arlington, VA ISBN 0–87355–098–6.

Every city-bound teacher should know this book. Nature's ability to triumph over asphalt and concrete permeates the text.

Science Books & Films. 1997. SB&F's Annual Science Book List. American Association for the Advancement of Science. Washington, DC. ISSN 0098–342X.

A selection of the best children's books reviewed in the 1996 volume year of *Science Books & Films*.

Seldin, Tim. 1996. At Home in the Natural World: The Montessori Approach to Science. Montessori Foundation. Alexandria, VA.

Full text available online at http://www.montessori.org.

Seller, Mick. 1993. Sound, Noise and Music. Glouster Press. ASIN 0–53117–408–5.

Good background information about sound as a form of energy, how it travels and is heard, and how music is made.

Smith, Susan Sperry. 1997. Early Childhood Mathematics. Allyn & Bacon. Needham Heights, MA. ISBN 0–20516–757–8.

National Council of Teachers of Mathematics standards-based book encourages teachers to create an active learning environment that fosters curiosity, confidence, and persistence in children as they learn math.

Spodek, Bernard, and Olivia N. Saracho. 1994. Right from the Start: Teaching Children Ages Three Through Eight. Allyn & Bacon. Needham Heights, MA. ISBN 0–20515–281–3.

Examines many issues concerning early childhood classrooms and learning, including science and mathematics for young children.

Sprung, Barbara, Merle Froschl, and Patricia B. Campbell. 1985. What Will Happen If...Young Children and the Scientific Method. Gryphon House. Beltsville, MD. ISBN 0–93162–902–0.

A guide for teachers to help incorporate math, science, and technology-related activities into the classroom in age-appropriate ways.

Vancleave, Janice. 1989. Biology for Every Kid: 101 Easy Experiments that Really Work. John Wiley & Sons. New York, NY. ISBN 0–47150–381–9.

Includes simple experiments with bean seedlings and molds.

Vancleave, Janice. 1991. Earth Science for Every Kid: 101 Easy Experiments that Really Work. John Wiley & Sons. New York, NY. ISBN 0–47154–389–6.

A broad array of Earth science and weather concepts are explored with experiments intended for the independent reader.

Vancleave, Janice. 1997. Janice Vancleave's Play and Find Out About Nature: Easy Experiments for Young Children. John Wiley & Sons. New York, NY. ISBN 0–47112–940–2.

Introduces students to plants and animals by having them create fun items. Provides students with opportunities to discover things on their own.

Van Rose, Susanna. 1994. Earth (Eyewitness Science). DK Publishing. London. ISBN 1–56458–476–3.

This *Eyewitness Science* book combines fine illustrations with concise nuggets of basic geological information about rock formation, earthquakes, volcanoes, erosion, and more, which makes it an intriguing reference volume.

Wagner, Sigrid (Ed.). 1993. Research Ideas for the Classroom: Early Childhood Mathematics, Vol. 1. NCTM. Reston, VA. ISBN 0–2897–287–2.

Volume one of a three-volume set. Contains interpretations of the research available on the teaching of mathematics to youngsters in early childhood and continuing through to students in high school.

Wicker, Lynn, and Dennis McKee. 1997. Simple and Fun Science. Essential Learning Products. Columbus, OH. ISBN 1–57110–171–3.

Part of the *Guided Practice* book series, this book contains information and activities about Earth and life sciences that reinforce science process skills.

Zim, Herbert, and Paul Shafer. 1989. Rocks and Minerals. Golden Books. New York. ISBN 0–60611–806–3.

Comprehensive pocket guide for identifying rocks.


## Early Childhood Education Policy


Elkind, David. 1995. Ties That Stress: The New Family Imbalance. Harvard University Press. Cambridge, MA. ISBN 0–67489–150–3.

Discusses what has happened to the American family in the last few decades and shows what the family has become.

Kagan, Sharon L. 1997. Leadership in Early Care and Education. NAEYC. Washington, DC. ISBN 0–93598–981–1.

Kagan, Sharon L. 1997. Not By Chance: Creating an Early Care and Education System for America's Children. Yale University. New Haven, CT. ISBN 0–93598–981–1.

Offers long-term solutions for improving the quality of early care and education of children under five years of age. The report calls for cooperative, cross-sector partnerships to build a system of high-quality child care.

Kagan, Sharon L., and Bernice Weissbourd. 1994. Putting Families First: America's Family Support Movement and the Challenge of Change. Jossey-Bass Publishers. San Francisco, CA. ISBN 1–55542–667–0.

Examines the evolution of current principles and practices in family support programs into mainstream institutions such as schools and the workplace. Discusses funding; local, state, and federal policies; and professional education.

Kagan, Sharon L., and Nancy E. Cohen (Eds.). 1996. Reinventing Early Care and Education: A Vision for a Quality System. Jossey-Bass Publishers. San Francisco, CA. ISBN 0–78790–319–1.

Provides a working blueprint for policy reform and program development. Examines issues related to providing high-quality services for young children, including parent engagement, licensing, professional development, regulation, governance, funding, and financing.

Miller, Patricia S., and James O. McDowelle. 1992. Administering Preschool Programs in Public Schools: A Practitioner's Handbook. Singular Publishing Group. San Diego, CA. ISBN 1–87910–578–0.

National Association for the Education of Young Children. 1996. NAEYC Position Statement on State Implementation of Welfare Reform. NAEYC. Washington, DC.

Full text available online at http://www.naeyc.org.

National Education Goals Panel. 1997. Special Early Childhood Report 1997. National Education Goals Panel. Washington, DC.

Pender, J. Anne, and Katherine Wrean (Eds.). 1993. Building Villages to Raise Our Children. Harvard Family Research Project. Cambridge, MA. ISBN 0–96306–271–9. http://www.hugse1.harvard.edu

Six guidebooks in this series offer advice on how to build a "village" in your community that links health, education, and social services.

Pipher, Mary. 1996. The Shelter of Each Other: Rebuilding Our Families. Putnam and Grosset Group. New York, NY. ISBN 0–39914–144–8.

Problems that families must confront are aggravated by the media and corporate values, the tools of technology, and demographic changes. Some families are cultivating qualities that will strengthen the bond between family members.

Scherer, Marge, et al. 1996. Working Constructively with Families. Educational Leadership, Vol. 53, No. 7. Association for Supervision and Curriculum Development. Alexandria, VA. ISSN 0013–1784.

A theme issue of *Educational Leadership*, this journal contains articles that examine several issues dealing with family and parental involvement in schooling.

Weiss, Heather B., et al. 1997. New Skills for New Schools: Preparing Teachers in Family Involvement. U.S. Department of Education. Washington, DC.

Full text available online at http://www.ed.gov/pubs

Zigler, Edward F., Sharon Lynn Kagan, and Nancy W. Hall (Eds.). 1996. Children, Families, and Government: Preparing for the Twenty-First Century. Cambridge University Press. Portchester, NY. ISBN 0–52158–940–1.

Provides analysis of the relationship between child development research and the design and implementation of social policy concerning children and families. Examines recent changes in our national ethos toward children and families.

## Programs and Research Centers

AIMS Education Foundation. Fresno, CA. P.O. Box 8120, Fresno, CA 93747. http://www.aimsedu.org

AIMS' mission is to integrate math and science learning for grades K–9.

Bank Street College of Education. New York, NY. 610 West 112th Street, New York, NY 10025. http://www.bnkst.edu

Bank Street College is composed of a graduate school of education, an elementary school for children, an on-site childcare center, a continuing education division, and a publication and media group. Bank Street College develops projects that meet the needs of families and children, including introducing new technologies into the classroom.

Center for Science and Mathematics Teaching. Tufts University. Medford, MA. 4 Colby Street, Medford, MA 02155. http://daffy.csmt.tufts.edu

CSMT's mission is to improve the teaching and learning of science in the nation's schools and universities by developing curricula, activities, and computer tools that allow students to actively participate in their own learning and to construct scientific knowledge for themselves.

Center on Families, Communities, Schools, and Children's Learning. Boston, MA. 605 Commonwealth Avenue, Boston, MA 02215.

Two research programs guide the Center's work: the Program on the Early Years of Childhood, which addresses issues facing children aged 0–10; and the Program on the Years of Early and Late Adolescence, which focuses on issues facing children aged 11–19.

Education Development Center, Inc. Newton, MA. 55 Chapel Street, Newton, MA 02158.

The Center's projects focus on improving and supporting science education reform nationwide. Projects range from instructional materials development to technical assistance.

Eisenhower National Clearinghouse for Mathematics and Science Education. Columbus, OH. 1929 Kenny Road, Columbus, OH 43210.

ERIC Clearinghouse on Elementary and Early Childhood Education (ERIC/EECE). Champaign, IL. 9 Children's Research Center, 51 Gerty Drive, Champaign, IL 61820. http://www.ericps.ed.uiuc.edu/eece

Offers information that focuses on early childhood and elementary school issues.

Even Start Family Literacy Program. Washington, DC. 600 Independence Avenue, SW, Washington, DC 20202 .http://www.ed.gov

This family-focused literacy program integrates early childhood education and adult education for parents.

Family Literacy Center and the ERIC Clearinghouse on Reading, English, and Communication. Bloomington, IN. 2805 East 10th Street, #150, Bloomington, IN 47408. http://www.ed.gov

Provides information about free and inexpensive literacy materials for parents, reviews and evaluates curricula and programs, and reports on research regarding literacy and families.

First Steps. Portsmouth, NH. 361 Hanover Street, Portsmouth, NH 03801. http://www.heinemann.com

First Steps is an international professional development and teaching resource developed to improve standards of literacy.

Harvard Family Research Project. Cambridge, MA. 38 Concord Avenue, Cambridge, MA 02138. http://www.hugse1.harvard.edu

Provides information on successful models for family-school-community partnerships, resources, and ideas about sustaining partnerships. Web site contains links to child and family research resources, child and family organizations, evaluation resources, government agencies and information, foundations, and more.

Head Start Program. Washington, DC. U.S. Department of Health and Human Services, Washington, DC 20201.

Head Start is a comprehensive program aimed at preschool children living in low-income families. It provides education as well as health screening and treatment, nutrition, and social services. Head Start actively seeks parental involvement.

Loyola University–Chicago. Erikson Institute. Chicago, IL. 820 N. Michigan Avenue, Chicago, IL 60611.

The Erikson Institute at the graduate school of Loyola University offers a Ph.D. in child development. It is designed to prepare practitioners and applied researchers for assuming leadership roles in their chosen fields and conducting competent independent research.

National Center for Family Literacy. Louisville, KY. 325 W. Main Street, Suite 200, Louisville, KY 40202. http://www.famlit.org

Advances and supports family literacy services through programming, training, research, advocacy, and dissemination of information about family literacy.

National Institute on Early Childhood Development and Education. Washington, DC. 555 New Jersey Avenue, NW, Washington, DC 20208. http://www.ed.gov

Promotes research, development, and dissemination of methods that will identify new approaches to improving young children's learning and development.

National Parent and Information Network. Champaign, IL. 51 Gerty Drive, Champaign, IL 61820. http://www.ed.gov

NPIN provides information to parents and to those who work with parents; it also fosters the exchange of parenting materials.

Teacher's College, Columbia University. Center for Young Children and Families. New York, NY. 525 W. 120th Street, New York, NY 10027. http://www.tc.columbia.edu

The goal of the Center is to conduct high-quality, policy-relevant, and interdisciplinary research on the development of children and families; to train young scholars and policy analysts; to provide information to those working directly with children and families; and to take a leadership role in national and state policy on children and families.

Wheelock College. The Centers for Child Care Policy and Training. Boston, MA. 200 The River Way, Boston, MA 02215.

The Centers include The Child Care Training Programs, The Family Child Care Project, and The Center for Career Development in Early Care and Education.

## Technology

Clements, D.H. 1992. Computers and Early Childhood Education. Advances in School Psychology: Preschool and Early Childhood Treatment Directions. Lawrence Erlbaum Associates. Mahwah, NJ. ISSN 0270–3920.

Clements, Douglas, Bonnie K. Nastasi, and Sudha Swaminathan. 1993. Research and Review. Young Children and Computers: Crossroads and Directions from Research. Young Children, Vol. 48, No. 2. Scranton, PA. ISSN 0044–0728.

Discusses the use of computers with young children.

Frazier, Max K. 1995. Caution: Students On Board the Internet. Educational Leadership Abstracts, Vol. 53, No. 2. Association for

Supervision and Curriculum Development. Alexandria, VA. ISSN 0013–1784.

Offers ways that teachers can protect students from negative influences on the Internet without discouraging their creative exploration.

Guthrie, Larry F., and Susan Richardson. 1995. Turned on to Language Arts: Computer Literacy in the Primary Grades. Educational Leadership Abstracts, Vol. 53, No. 2. Association for Supervision and Curriculum Development. Alexandria, VA. ISSN 0013–1784.

The authors report on visits to more than 50 classrooms nationwide, concluding that educational technology is changing how teachers teach and students learn.

Papert, Seymour. 1993. The Children's Machine: Rethinking School in the Age of the Computer. Basic Books. Scranton, PA. ISBN 0–46501–830–0.

Peha, Jon M. 1995. How K–12 Teachers Are Using Computer Networks. Educational Leadership Abstracts, Vol. 53, No. 2. Association for Supervision and Curriculum Development. Alexandria, VA. ISSN 0013–1784.

Findings from a Carnegie Mellon University study, conducted in conjunction with the Pittsburgh Public Schools. Through classroom interviews with Internet users, the study's researchers offer present findings and future projections for technology use. The author suggests ways to overcome obstacles, gives advice on providing effective staff development, and shares tips for reducing installation costs.

Raphael, Jacqueline, and Richard Greenberg. 1995. Image Processing: A State-of-the-Art Way to Learn Math and Science. Educational Leadership Abstracts, Vol. 53, No. 2. Association for Supervision and Curriculum Development. Alexandria, VA. ISSN 0013–1784.

Compares how students can obtain the same benefits from image processing as scientists who are on the cutting edge of research.

Van Dusen, Lani M., and Blaine R. Worthen. 1995. Can Integrated Instructional Technology Transform the Classroom? Educational Leadership Abstracts, Vol. 53, No. 2. Association for Supervision and Curriculum Development. Alexandria, VA. ISSN 0013–1784.

An in-depth look at Computer-Based Integrated Learning Systems (ILS), their potential, and their drawbacks. It also offers guidelines on how to make ILS more useful.

Wright, J.L., and D.D. Shade (Eds.). 1994. Young Children: Active Learners in a Technological Age. NAEYC. Washington, DC. ISBN 0–93598–963–3.

## Web Sites for Teachers and Families

Allison, Jane. Computers in Elementary Education. http://nimbus.temple.edu/~jallis00/

Designed for the elementary school computer teacher. Provides Internet links to sites that can be used when creating classroom projects in all subjects at the elementary school level.

The Annenburg CPB Math and Science Project. Mathematics Learning Forums. http://www.pbs.org/teacher/math

Provides online seminars to elementary and middle school teachers to help them refine their teaching practice and to encourage their interest in the math standards of the National Council of Teachers of Mathematics.

Cool Sites for Kids. http://www.ala.org/parents/index.html

For homework help, reference tools, or fun. Selected by the American Library Association.

The ERIC Review. ERIC Clearinghouse on Elementary and Early Childhood Education. Champaign, IL. http://ericeece.org

Provides education practitioners with research and news; announces important ERIC developments, new products, and services; and presents recent research findings.

Franklin Institute Science Museum. http://sln.fi.edu

From the Philadelphia-based science center. For families and the general public.

Internet Public Library. http://www.ipl.org

Links to sites and reference materials. Also lets viewers ask reference questions.

Katz, Lilian G. (Ed.). 1998. Early Childhood Research and Practice. http://ecrp.uiuc.edu

Peer-reviewed electronic journal covers topics related to growth, learning, development, care, and education of children through age 8. Includes articles, abstracts, and essays.

Kids Only: Internet Resources. http://vm.sc.edu/~beaulib/kidsonly.html

Mega-Mathematics. http://www.cs.uidaho.edu

Problems and explanations, games, and challenges.

National Institute for Family Literacy. http://www.nifl.gov

NIFL provides an online discussion group for family literacy issues. To participate in this discussion, send a message to LISTPROC@NOVEL.NIFL.GOV saying: subscribe NIFL-Family first_name last_name. Spell your first and last names as you would like them to appear. Do not include any other text in the message.

National Museum of the American Indian. http://www.logomancy.com/heye.html

Provides educational information for all ages from the newest Smithsonian museum.

Scholastic. http://www.scholastic.com

Yahooligans. http://www.yahooligans.com

A search engine designed specifically for children and young adults. Search subjects or topics such as homework help, countries, and science.

# Forum on Early Childhood Science, Mathematics, and Technology Education: Policy, Partners, and Practice February 6–8, 1998

## List of Forum Attendees

Sandi Abell
Purdue University

Olaiya Aina
University of Charleston

Sally Anderson
Vermont Center for the Book

Zoe A. Barley
Western Michigan University

Kathy Beasley
Averill Elementary School
Lansing, MI

Daniel B. Berch
NIH/NICHD
Bethesda, MD

Marilyn Berman
Science Linkages in the Community
Rochester, NY

Bennett I. Bertenthal
National Science Foundation

Lowell Bethel
National Science Foundation

Eve M. Bither
Department of Education/ OERI
Washington, DC

Barbara T. Bowman
Erikson Institute
Chicago, IL

John S. Bradley
National Science Foundation

Cathy Bradshaw
AAAS Black Church Initiative
Dorchester, MA

Martha Bronson
Boston College

Karen Brown
Madison Avenue Family Life Center
Baltimore, MD

Sherry P. Brown
J. Wallace James Elementary
Lafayette, LA

Barbara H. Butler
National Science Foundation

Alverna Champion
National Science Foundation

Edward Chittenden
Educational Testing Service
Princeton, NJ

Ines L. Cifuentes
Carnegie Institution of Washington

Douglas H. Clements
State University of New York–Buffalo
Williamsville, NY

Grace Davila Coates
University of California–Berkeley

Michael Cohen
Domestic Policy Council
Washington, DC

Judi Colombaro
Please Touch Museum
Philadelphia, PA

Juanita V. Copley
University of Houston

Carol Copple
National Association for the Education of Young Children
Washington, DC

Margaret B. Cozzens
National Science Foundation

Barbara Day
The University of North Carolina

Debby Deal
George Mason University

Judy DeLoache
University of Illinois

Rheta DeVries
University of Northern Iowa

Diane Trister-Dodge
Teaching Strategies, Inc.
Washington, DC

Carolyn Pope Edwards
University of Nebraska

Lois Edwards
Minneapolis, MN

Linda Espinosa
University of Missouri–Columbia

Joyce B. Evans
National Science Foundation

Greta G. Fein
University of Maryland

Francis M. Fennell
National Science Foundation

Hyman H. Field
National Science Foundation

Gail Foster
BSCS
Colorado Springs, CO

Judd D. Freeman
National Science Foundation

Lucia French
University of Rochester

Carol Sue Fromboluti
Department of Education/OERI
Washington, DC

Barbara Gilkey
AR HIPPY
Little Rock, AR

Carole Greenes
Boston University

Aimee Guidera
National Alliance of Business
Washington, DC

Dominic F. Gullo
University of Wisconsin–Milwaukee

Cheryl Hall
Easter Seals
Calverton, MD

Michael R. Haney
National Science Foundation

Jean D. Harlan
Clinical Psychologist
Oak Creek, WI

Charles Hohmann
High/Scope Educational Research Foundation
Ypsilanti, MI

Kimi Hosoume
University of California–Berkeley
Berkeley, CA

Daniel Householder
National Science Foundation

John Hunt
National Science Foundation

John Hunter
Austin State

Patricia Hutchinson
The College of New Jersey

Tabitha A. Ishmael
Howard University Early Learning Programs
Washington, DC

Charles C. James
Carnegie Institution of Washington

Renee Jefferson

Fred Johnson
Shelby County Board of Education
Memphis, TN

Jacqueline R. Johnson
Grand Valley State University
Allendale, MI

Valorie Johnson
W. K. Kellogg Foundation

Donald E. Jones
National Science Foundation

Jacqueline Jones
Educational Testing Service

Vinetta Jones
College Board
Washington, DC

Wendy Jones
The Country Day School
McLean, VA

Naomi Karp
Department of Education
Washington, DC

Lilian Katz
ERIC Clearinghouse on
Elementary and Early
Childhood Education
Champaign, IL

Patricia F. Kinney
National Science Foundation

Elon Kohlberg
Harvard University

Carole B. Lacampagne
Department of Education
Washington, DC

Twila Liggett
The Reading Rainbow

Karen K. Lind
University of Louisville

Mary M. Lindquist
College of Education

Faite R.P. Mack
Grand Valley State University
Grand Rapids, MI

Shirley Malcom
American Association for the
Advancement of Science

Rebecca A. Marcon
University of Northern Florida

Cecille Martinez-Spall
Santa Fe Public Schools

Christine Massey
University of Pennsylvania

Dorothy McCormick
The Country Day School
McLean, VA

Donald Miller
Oakland University
Rochester, MI

Jan Morrison
The Park School
Brooklandville, MD

Patricia M. Morse
National Science Foundation

Fritz Mosher
Carnegie Corporation of
New York

Gene Myers
Western Washington
University
Bellingham, WA

Christine B. Neely
Audubon Zoo
New Orleans, LA

Greg Nelson
Pacific Lutheran University
Tacoma, WA

Rebecca S. New
University of New Hampshire

Deborah J. Norris
Oklahoma State University

Pat O'Connell Ross
OERI/U.S. Department of
Education

James Oglesby
National Science Foundation

Yolanda N. Padrón
University of Houston

Sandra Parker
American Association for the
Advancement of Science

Deborah Phillips
National Research Council

Catherine Prevour
Trinity College
Washington, DC

Mary Rivkin
University of Maryland–
Baltimore

Nanette Roberts
Scarsdale, NY

Vanessa Robinson
Mabel Barrett Fitzgerald
Daycare Center
New York, NY

Samuel Rodriguez
Department of Energy

Margie Rosario
Science Linkages in the
Community
Rapid City, SD

Anastasia Samaras
The Catholic University
of America
Washington, DC

Beverly S. Sanford
Science Works

Joel Schneider
Children's TV Workshop
New York, NY

Robin Sharp
San Francisco Community
School

Maxine F. Singer
Carnegie Institution of
Washington

Deborah C. Smith
Michigan State University

Susan Sperry Smith
Cardinal Stritch University
Milwaukee, WI

Susan P. Snyder
National Science Foundation

Amanda Sonoma
National Alliance of Business
Washington, DC

Prentice Starkey
University of California–
Berkeley

Keiko Takayama
Washington, DC

Lisa Tassone

Bernida Thompson
Roots Activity Learning
Center
Washington, DC

Rosemarie Truglio
Children's Television
Workshop

Jan Tuomi
National Academy of Science

David P. Weikart
High/Scope Educational
Research Foundation
Ypsilanti, MI

Molly Weinburgh
Georgia State University
Atlanta, GA

Luther S. Williams
National Science Foundation

Virginia Williams
National Council of Teachers
of Mathematics

Ruth Wilson
Bowling Green State
University

June L. Wright
Computer Discovery Project
Willimantic, CT

Emily Wurtz
National Education Goals
Panel

Judy Wurtzel
U.S. Department of Education

## Project 2061 Staff

George D. Nelson
Director

Andrew Ahlgren
Associate Director

Gerald Kulm
Program Director

Jo Ellen Roseman
Curriculum Director

Mary Koppal
Communications Director

Mary Ann Brearton
Field Services Coordinator

Lester Matlock
Project Administrator

Soren Wheeler
Project Assistant

Cheryl Wilkins
Secretary

Keran Noel-Tarpley
Publications Assistant

## Index

WITHDRAWN

GOSHEN COLL E - GOOD LIBRARY
3 9310 01034621 9